BLUE FUTURE

ALSO BY THE AUTHOR

Parcel of Rogues: How Free Trade Is Failing Canada

Take Back the Nation (with Bruce Campbell)

Take Back the Nation 2 (with Bruce Campbell)

Class Warfare: The Assault on Canada's Schools
(with Heather-Jane Robertson)

Straight through the Heart: How the Liberals Abandoned the Just Society
(with Bruce Campbell)

The Big Black Book: The Essential Views of Conrad and
Barbara Amiel Black (with Jim Winter)

MAI: The Multilateral Agreement on Investment and the Threat to Canadian
Sovereignty (with Tony Clarke)

MAI: The Multilateral Agreement on Investment and the Threat to American
Freedom (with Tony Clarke)

The Fight of My Life: Confessions of an Unrepentant Canadian

MAI: The Multilateral Agreement on Investment Round 2; New Global and
Internal Threats to Canadian Sovereignty (with Tony Clarke)

Frederick Street: Life and Death on Canada's Love Canal
(with Elizabeth May)

Global Showdown: How the New Activists Are Fighting Global Corporate
Rule (with Tony Clarke)

Blue Gold: The Battle Against Corporate Theft of the World's Water
(with Tony Clarke)

Profit Is Not the Cure: A Citizen's Guide to Saving Medicare

Too Close for Comfort; Canada's Future Within Fortress North America

Blue Covenant: The Global Water Crisis and the Fight for the Right to Water

BLUE FUTURE

PROTECTING WATER
FOR PEOPLE AND THE
PLANET FOREVER

MAUDE
BARLOW

THE NEW PRESS

NEW YORK

Requests for permission to reproduce selections from this book should be mailed to: Permissions Department, The New Press, 120 Wall Street, 31st floor, New York, NY 10005.

First published in Canada by House of Anansi Press, Inc., Toronto, 2013
Published in the United States by The New Press, New York, 2014
Distributed by Perseus Distribution

Library-of-Congress Cataloging-in-Publication Data

Barlow, Maude.
  Blue future : protecting water for people and the planet forever / Maude Barlow.
    pages cm.
  Includes bibliographical references and index.
  ISBN 978-1-59558-947-7 (hardback)—ISBN 978-1-59558-948-4 (e-book) 1. Water resources development—Economic aspects. 2. Water resources development—Government policy. 3. Water rights—Government policy. 4. Water-supply—Political aspects. 5. Water-supply—Economic aspects. I. Title.
  HD1691.B3665 2014
  333.91'116—dc23

                                                                2013036341

The New Press publishes books that promote and enrich public discussion and understanding of the issues vital to our democracy and to a more equitable world. These books are made possible by the enthusiasm of our readers; the support of a committed group of donors, large and small; the collaboration of our many partners in the independent media and the not-for-profit sector; booksellers, who often hand-sell New Press books; librarians; and above all by our authors.

www.thenewpress.com

Book design and composition by Alysia Shewchuk
This book was set in Minion and Frutiger

Printed in the United States

10 9 8 7 6 5 4 3 2 1

To Miguel d'Escoto Brockmann and Pablo Solón,
who never lost the belief that we could make
the right to water real

CONTENTS

Our world in stupor lies;
Yet, dotted everywhere,
Ironic points of light
Flash out wherever the Just
Exchange their messages

— W. H. Auden, "September 1, 1939"

INTRODUCTION

O N JULY 28, 2010, the United Nations General Assembly
adopted an historic resolution recognizing the human right to
safe and clean drinking water and sanitation as "essential for the full
enjoyment of the right to life." For those of us in the balcony of the
General Assembly that day, the air was tense. A number of powerful
countries had lined up to oppose it, so it had to be put to a vote. The
Bolivian ambassador to the UN, Pablo Solón, introduced the resolution by reminding the assembly that humans are composed of about
two-thirds water and that our blood flows like a network of rivers to
transport nutrients and energy through our bodies. "Water is life,"
he said.

Then he laid out the story of the number of people around the
world who were dying from lack of access to clean water and quoted
a new World Health Organization study on diarrhea showing that,
every three and a half seconds in the developing world, a child dies
of waterborne disease. Ambassador Solón then quietly snapped his
fingers three times and held his small finger up for a half-second. The
General Assembly of the United Nations fell silent. Moments later, it
voted overwhelmingly to recognize the human rights to water and
sanitation. The floor erupted in cheers.

The recognition by the General Assembly of these rights represented a breakthrough in the struggle for water justice in the world.

It followed years of hard work and was a key platform of our global water justice movement for at least two decades. For me personally, it was the culmination of many years of work, and I was proud and grateful to all who had helped make it happen.

But our work is far from over. Recognizing a right is simply the first step in making it a reality for the millions who are living in the shadow of the greatest crisis of our era. With our insatiable demand for water, we are creating the perfect storm for an unprecedented world water crisis: a rising population and an unrelenting demand for water by industry, agriculture, and the developed world; over-extraction of water from the world's finite water stock; climate change, spreading drought; and income disparity between and within countries, with the greatest burden of the race for water falling on the poor.

"Suddenly it is so clear: the world is running out of fresh water." These were the opening words of my 2002 book, *Blue Gold: The Battle Against Corporate Theft of the World's Water* (co-written with Tony Clarke), which warned of a mighty contest brewing over the world's dwindling freshwater supplies. As water became the oil of the twenty-first century, we predicted, a water cartel would emerge to lay claim to the planet's freshwater resources. This has come true. But so has our prediction that a global water justice movement would emerge to challenge the "lords of water."

In my 2007 book, *Blue Covenant: The Global Water Crisis and the Coming Battle for the Right to Water*, I described the growing water cartel and its relentless drive to find ways to take control of the world's water supplies. I also reported on the amazing work of the environmentalists, human rights activists, indigenous and women's groups, small farmers, peasants, and thousands of grassroots communities that make up the global water justice movement fighting for the right to water and to keep water under public and democratic control.

In the six years since *Blue Covenant* was published, much has been accomplished. Reports on the crisis are commonplace in mainstream media and the classroom. Books, films, and music move millions to action. The United Nations, other global institutions, and many universities are also sounding the alarm. A movement has coalesced to provide water and sanitation to the urban and rural poor, with mixed, but hopeful, results.

Yet in those same years the water crisis dramatically deepened. It is now accepted that, with the unexpected growth in both population and new consumer classes in almost every country, global demand for water in 2030 will outstrip supply by 40 percent. A report from the U.S. global intelligence agencies warns that one-third of the world's people will live in basins where the deficit is more than 50 percent. Five hundred scientists from around the world met in Bonn in May 2013 at the invitation of UN Secretary-General Ban Ki-moon and sent out a warning that our abuse of water has caused the planet to enter a "new geologic age." They likened this "planetary transformation" to the retreat of the glaciers more than 11,000 years ago. Within the space of two generations, the majority of people on the planet will face serious water shortages and the world's water systems will reach a tipping point that could trigger irreversible change, with potentially catastrophic consequences. Already, the world-renowned scientists said, a majority of the world's people live within 50 kilometres of an impaired water source — one that is running dry or polluted.

The stage is being set for drought on an unprecedented scale, for mass starvation and the migration of millions of water refugees leaving parched lands to look for water. All the justice and awareness in the world cannot stave off this future if the water is not there.

Open any textbook on water and you will see the numbers: how many children die every day; where the water tables have dried up; how aquifers are being depleted. Yet we continue to extract from our

precious rivers and lakes and pump our groundwater, using the last
of a finite supply of water that will be needed if future generations
and other species are to survive.

Amazingly, most of our political leaders ignore the water crisis
and create policy decisions as if there were no end to water supplies.
They continue to be captives of an economic framework that promotes unlimited growth, unregulated trade, and bigger and more
powerful (and increasingly self-governing) transnational corporations, all of which hasten the destruction of our supplies of fresh
water. Somewhere between the hard truths about the world's water
crisis and this perplexing denial on the part of political and corporate leaders, millions — soon to be billions — struggle to deal with
disappearing watersheds.

The story does not need to end in tragedy. There are solutions to
our water crisis and a path to a just and water-secure world. To get
to this place, however, we must establish principles to guide us and
help us create policies, laws, and international agreements to protect
water and water justice, now and forever.

This book puts forward four principles for a water-secure future.
Principle one, "Water Is a Human Right," addresses the current reality of water inequality and lays out a road map to fixing the problem. Principle two, "Water Is a Common Heritage," argues that water
is not like running shoes or cars and must not be allowed to become
a commodity to be bought and sold on the open market. Principle
three, "Water Has Rights Too," makes the case for protection of
source water and watershed governance and the need to make our
human laws compatible with those of nature if we are to survive. The
fourth principle, "Water Can Teach Us How to Live Together," is a cry
from the heart to come together around a common threat — the end
of clean water — and find a way to live more lightly on this planet.

The grab for the planet's dwindling resources is the defining
issue of our time. Water is not a resource put here solely for our

convenience, pleasure, and profit; it is the source of all life. It is urgent that we clarify the values and principles needed to protect the planet's fresh water. I offer this book as a guide.

WATER IS A HUMAN RIGHT

This principle recognizes that denying people or communities access to drinking water and sanitation is a violation of their human rights. In our world today, wealthy people and corporations have access to all the water they want while millions go without because they cannot pay for it or do not have access to it. The right to water is not free for all, allowing anyone to use all they want for any purpose; rather, it guarantees clean, accessible drinking water and sanitation for personal and domestic use for all. The human right to water places the onus on governments to provide water and sanitation to their people and to prevent harm to the source waters that supply it. Most essentially, the human right to water is an issue of justice, not charity. It requires a challenge to the current power structures that support unequal access to the world's dwindling freshwater supplies.

1

THE CASE FOR
THE RIGHT TO WATER

Small battles are being won around the world, but I think people
are losing. I do see the present and the future of our children as
very dark. But I trust the people's capacity for reflection, rage
and rebellion. —Oscar Olivera, leader of the Cochabamba water
revolution[1]

E VERY YEAR, MORE PEOPLE die from unsafe water than from
all forms of violence, including war.

Some 3.6 million people, 1.5 million of whom are children, die
every year from water-related diseases, including diarrhea, typhoid,
cholera, and dysentery. One billion people still defecate in the open,
and 2.5 billion live without basic sanitation services. By 2030 more
than 5 billion people — nearly 70 percent of the world's population —
may be without adequate sanitation.

Living without clean water and sanitation has enormous ramifications for both families and societies. It is always hardest on
the women and children. The United Nations reports that women
spend about 40 billion hours collecting water every year. In many
countries, women spend as much as five or six hours a day fetching

water, and their female children accompany them, thereby losing the opportunity to go to school. According to the 2012 UN report on the Millennium Development Goals, women in sub-Saharan Africa spend a collective average of 200 million hours per day gathering water, and more than two-thirds of the burden for water and sanitation falls on women and girls.

Many girls also do not attend school because there are no private toilet facilities for them to use. Amnesty International says that the right to sanitation

> means that people should not be left with no option but to defecate in the open, or into a bucket or a plastic bag. Women and girls should not have to choose between going to a public toilet or risking sexual violence. They should not — due to lack of toilets in schools — be forced to choose between education and dignity. Children should not be in a situation where lack of an adequate toilet or lack of information about safe hygiene puts them at risk of death from diarrhoea.[2]

WATER INJUSTICE

In every case, if these families had the means, their children would not be dying and would be attending school. The lack of access to clean water and sanitation, in terms of sheer numbers affected, is arguably the single most urgent human rights issue of our time.

Most in danger are those living in slums or impoverished rural communities in Latin America, Asia, and Africa. Peri-urban slums ring most of the developing world's cities, where climate and food refugees are arriving in relentless numbers. Unable to access their traditional sources of water because they have disappeared or been polluted, and unable to afford the high rates set by newly privatized water services, these refugees must rely on drinking water sources contaminated by their own untreated human waste as well as industrial poisons.

The growing commodification of the world's water has made it increasingly inaccessible to those without money. Many poor countries have been strongly encouraged by the World Bank to contract water services to private for-profit utilities, a practice that has spawned fierce resistance by the millions left out because of poverty. Other struggles are taking place with bottled-water companies that drain local water supplies. There are "land grabs" in which countries and investment funds buy up massive amounts of land in the Global South for access to the water and soil at a future time.

Some countries actually auction off water to global interests such as mining companies, which now literally own the water that used to belong to everyone. And many countries are introducing water markets and water trading, whereby water licences — often owned by private companies or industrial agribusiness — are allowed to be hoarded, bought, sold, and traded, sometimes on the international open market, to those that can afford to buy it. In all of these cases, water becomes the private property of those with the means to buy it and is increasingly denied to those without. All over the world, private citizens, small farmers, peasants, indigenous people, and the poor have found themselves unable to stand up to these corporate interests.

The victims are more likely to come from developing countries. By every measurement, global income disparities are the severest they have been in almost a century, with a small percentage of the world's elite owning the vast majority of its assets. In a January 2013 report, Oxfam International says that the explosion in extreme wealth and income is exacerbating inequality and hindering the world's ability to tackle poverty. The $240 billion net 2012 income of the hundred richest billionaires would be enough to make extreme poverty history four times over. The richest 1 percent has increased its income by 60 percent in the past twenty years, reports Oxfam, with the financial crisis accelerating rather than slowing the process.

Oxfam warns that extreme wealth and income are economically inefficient, politically corrosive, socially divisive, and environmentally destructive. "Concentration of resources in the hands of the top one per cent depresses economic activity and makes life harder for everyone else — particularly those at the bottom of the economic ladder," says executive director Jeremy Hobbs. "In a world where even basic resources such as land and water are increasingly scarce, we cannot afford to concentrate assets in the hands of a few and leave the many to struggle over what's left," he adds.[3]

And yet asset concentration is real. Billions around the world live in poverty amongst great wealth, and this negatively affects their access to water. A child born in the northern hemisphere consumes thirty to fifty times as much water as one in the southern. Per capita daily water use in North America and Japan is 350 litres, in Europe it is 200 litres, and in sub-Saharan Africa it is 10 to 20 litres. An estimated 90 percent of the three billion people expected to be added to the population by 2050 will be from the developing world.[4]

But the crisis is not limited to people who live in the Global South. As we see deepening income inequality in the countries of the First World, water cut-offs are now happening to the poor there too. Tens of thousands of inner-city residents in Detroit, Michigan, have no running water because they cannot afford the rising tariffs. Unemployment in the affected communities runs at about 50 percent. Residents are forced to run hoses from neighbouring homes or take water canisters to public washrooms for fill-ups. Social services have removed children from some homes, citing lack of running water. Cut-offs are also taking place in Europe, where recent austerity measures are driving the cost of basic necessities beyond the ability of many to pay.

Nor is all the water wealth in the rich countries of the North. Renowned Indian movie director Shekhar Kapur fears for the future of his beloved country, in which such extremes of wealth and poverty coexist. He writes:

> In Mumbai. Just across the road from Juhu Vile Parle Scheme, all the beautiful people and film stars live opposite a slum called Nehru Nagar. Once a day, or maybe even less, water arrives in tankers run by the local "water mafia" and their goons. Women and children wait in line for a bucket of water, and fights break out as the tankers begin to run dry.
>
> Yet, literally across the road, the "stars" after their workouts in the gym or a day on a film set can stay in the shower for hours. The water will not stop flowing. Often at less than half the cost that the slum dwellers pay for a single bucket of water.[5]

In many countries the rich can access all the water that money can buy, while the poor — usually women and children — walk kilometres to find water that may or may not be clean enough to drink. In many poor countries, tourists and the wealthy have preferential access to clean water for resorts, golf courses, and spas while local slums have no running water. Millions live in "informal settlements" unrecognized by governments, which consequently do not provide basic services to their inhabitants.

RUNNING OUT

As a rule, poverty and class divisions are at the root of lack of access to clean water. But increasingly the crisis is due as well to a decline in local water sources that in turn forces people to become refugees. Over-extraction of water for industrial food production, so-called

economic development, and water-reliant natural resource extraction is taking a terrible toll on the world's finite freshwater supplies.

The lesson we all learned as children — that we cannot run out of
water because of the endless workings of the hydrologic cycle — is simply not true. While the water is still on the planet somewhere, because
of our engineering of the world's water supplies to promote industrial
development, it is not drinkable or in the right place. As a result, many
communities are running out of accessible clean water. We humans
are polluting, mismanaging, and displacing water at an alarming rate.

Global water withdrawals have risen 50 percent in the past several decades and are still increasing dramatically. Using bore-well
technology that did not exist a hundred years ago, humans are now
relentlessly mining groundwater. Worldwide pumping of groundwater more than doubled between 1960 and 2000 and is responsible
for about 25 percent of the rise in sea levels.[6] By 2030 it is expected
that demand will outstrip supply by 40 percent and almost half the
world's population will be living in areas of high water stress. By 2075
the number affected could be as high as seven billion.[7]

This increase in demand is due to a combination of industrial-
ization, exponential population growth, and more people leading
a water-intensive consumer lifestyle. The demand for water is insatiable on a planet whose population is approaching 9 or 10 billion
people by 2050. To house its population, in the next two decades
China alone is planning to build 500 new cities with more than
100,000 people each. India will add 600 million people to its population by 2050, giving it the highest population in the world. Pakistan
will be approaching 300 million, Nigeria 290 million, and Uganda
93 million. Malawi even now cannot feed its population of 13 million;
by 2050 an estimated 32 million people will be living there.[8]

Peter Gleick, an American scientist who founded the Pacific
Institute, which does pioneering research on water and climate,
reminds us that, while the population is growing, the amount of

accessible water is finite. In 1950 the population of the United States was 150 million; it is more than 315 million today. Jordan had a million people in 1960; it has 6 million today. Iraq had around 7 million in 1960, and today its population exceeds 31 million. All these new populations must share finite water supplies that were being consumed by much smaller populations decades ago.[9]

In their book *Out of Water*, Colin Chartres and Samyuktha Varma estimate the growth in our per capita use of water globally in relation to population growth. If we include the water used to grow our food (known as virtual water), then a person who consumes 2,500 calories per day will consume 2,500 litres of water. Multiplied by 365 days per year, this totals almost 100 million litres — one megalitre — per person. If the population grows to 9 billion by 2050 (most figures predict it will be higher), the water needed will equal the capacity of at least another twenty-five to fifty enormous dams similar to the Aswan High Dam on the Nile River in Egypt. The authors point out that these vast amounts of water are simply not available, or at least not available in the areas where we need them to produce food.[10]

A study from the University of Twente in the Netherlands puts the average global footprint much higher. Virtual water expert Professor Arjen Hoekstra reports that if all the water used for our daily lives is factored in, the average per person daily water consumption is 4,000 litres.[11]

Of course, the way we live determines how much water we use and abuse. Almost half the world's population is still living on the land, much as in previous generations, sustainably using and caring for local water sources. That means the rest are using far more than their share. For instance, global meat production is predicted to double by 2050, using 70 percent of all agricultural land and consuming one-third of the world's grain. The rich consume most of this: people in the wealthy North consume three times as much meat and four times as much milk as people in the South.

Writer and journalist George Monbiot wrote in the *Guardian* that the economy is growing much faster than the rate of population and that economic growth is the real threat. Global consumption will increase so much that by the end of the twenty-first century we will have used sixteen times more economic resources than humans have consumed since we "came down from the trees," says Monbiot.[12] Yet it is the mantra of governments almost everywhere to "grow" their way to prosperity, putting the world's water supplies at grave risk.

Already we are seeing the results of overexploitation. The world's rivers — the single largest renewable water resource and a crucible of aquatic biodiversity — are in crisis from pollution and over-extraction. About 1.4 billion people live in river basins where all the blue water (fresh surface and groundwater) is already committed or overcommitted. The journal *Nature* reports that nearly 80 percent of the world's human population lives in areas where river waters are highly threatened, posing a major threat to human water activity.[13]

Desertification is advancing rapidly in more than a hundred countries, through over-extraction of rivers and groundwater and the advance of climate change, sending millions of refugees in search of safe haven. The Earth Policy Institute's Lester Brown, an influential American writer and environmentalist who founded the Worldwatch Institute, reports that the Sahara Desert is expanding in every direction, squeezing the populations of Tunisia, Morocco, and Algeria. The Sahelian swath of savannah that separates the southern Sahara from the tropical rainforests of central Africa is shrinking and the desert is moving south, invading populous Nigeria. Lake Chad, once the sixth largest lake in the world, is 90 percent gone, putting the lives and livelihoods of 30 million West Africans in danger.[14]

Some 600,000 square kilometres of land are now desert in Brazil, and Mexico is forced to abandon 250,000 square kilometres of farmland to desert every year. Their rural refugees gravitate to the slums of Buenos Aires, São Paulo, and Mexico City. Dr. Kevin Trenberth, who is with the World Climate Research Programme of the United Nations, projects that by 2055, between 80 and 170 million people in Latin America will likely have insufficient water for their basic needs.[15]

Hundreds of thousands of "environmental refugees" have had to flee their homes in central Asia as the Aral Sea, once the fourth largest lake in the world, dies because of massive cotton irrigation during the years of the Soviet Union. Iran's Lake Urmia, the largest lake in the Middle East and the third largest salt lake in the world, is 60 percent gone and may dry up completely. The lake used to provide crop irrigation and fish for the tens of millions who live within a few hundred kilometres of the lake, but drought has increased its salinity to levels too high to provide either anymore.

Over the past half-century, some 24,000 villages in northern and western China have been abandoned entirely or partially because of desert expansion. (An additional 450 "cancer villages" have been identified for evacuation.) Lester Brown says that China is heading for a "dust bowl" that could force migration that might number in the tens of millions.[16]

These conditions are not limited to countries in the south. During the heat-scorched summer of 2012 in eastern Canada and the United States, a new study by a group of American scientists, published in the journal *Nature Geoscience*, stated that the drought experienced by western North America during the past decade is the worst in eight hundred years. The situation will decline steadily, say the authors, and the droughts we are experiencing now will likely be seen as the "wet" end of a drier hydroclimate predicted for the rest of the twenty-first century.[17]

The Ogallala Aquifer, the once mighty underground lake that runs from the eastern slope of the Rockies to the Texas Panhandle and has provided water for America's breadbasket, is running out. "The Ogallala supply is going to run out and the Plains will become

uneconomical to farm," says David Brauer of the Ogallala Research
Service, an agency of the U.S. government's agriculture department.
"That is beyond reasonable argument. Our goal now is to engineer a
soft landing. That's all we can do."[18]

If water takings from the Great Lakes of North America are similar to those of global groundwater takings, the Great Lakes could be
bone-dry in eighty years, says Marc Bierkens, professor of hydrology
at Utrecht University and principal author of a groundbreaking 2010
global study on groundwater takings. He says the size of the global
groundwater footprint — the area required to sustain groundwater
use and groundwater-dependent ecosystem services — is currently
3.5 times the actual area of aquifers, and that about 1.7 billion people
live in areas where groundwater resources and/or groundwater-
dependent ecosystems are under threat.[19]

The 2011 and 2012 droughts in Europe were the worst in a hundred years, with withered crops and shrinking rivers and lakes
becoming commonplace. The Mediterranean is particularly hard
hit. Groundwater has fallen 80 percent in Italy's Milan district, and
in Turkey, Lake Akşehir — once three times the size of Washington,
DC — has disappeared. The World Wildlife Fund reports that more
than 50 percent of the wetlands of the Mediterranean have dried up
and that a land area the size of the United Kingdom is under threat
of desertification.[20]

WATER REFUGEES

While the North American and European crises may not produce as
many internal water refugees as some other parts of the world, these
regions will be asked to open their doors to water refugees. They will
be seen as destinations for millions, possibly billions, of water refugees from the Global South. A UN conference on desertification in

Tunisia projected that by 2020 up to 60 million people might have migrated from sub-Saharan Africa to North Africa and Europe. Another United Nations study predicts that 2.2 million migrants will arrive in the rich world every year, from now till 2050. Britain's population will rise by almost 16 million, almost all from migration. The UN's population division says this migration will bring about an upheaval without parallel in human history.[21]

Some experts say the population bomb will peak and will not necessarily lead to environmental devastation. The UN's prediction of 9 or 10 billion by 2050 likely represents the end of population growth, says British environmental writer Fred Pearce. In his book *The Coming Population Crash*, Pearce documents the fact that women almost everywhere are having fewer children: half the number their mothers did, in fact. He attributes this trend to education and the empowerment of women. Within the next two generations, he says, this will lead to lowered world fertility rates and a return to more sustainable populations. Pearce agrees that there will be an increase in migration rates, but he says they will be largely for the good. Countries with low birth rates need young people and new workers, and migration can be a win-win situation if planned and done sustainably.[22]

While this is good news, it is crucial to preserve the earth's resources through this period of intense growth in demand and to share them more equitably. No place on earth will be free from the consequences of the water crisis now unfolding. Even if we start to slow the damage we have created, by challenging the growth imperative and adopting water conservation practices and source-water protection, it is crucial that we set rules of fairness and justice around the issue of access. Otherwise we will increasingly see a world deeply divided between those with access to clean water and those without — literally a world divided by the right to live.

2

THE FIGHT FOR
THE RIGHT TO WATER

Is access to water a human right or just a need? Is water a common good like air or a commodity like Coca-Cola? Who is being given the right or the power to turn the tap on or off—people, governments or the invisible hand of the market? Who sets the price for a poor district in Manila or La Paz—the locally elected water board or the CEO of a transnational water corporation in another country?—**Rosmarie Bär, Alliance Sud, Switzerland**

WATER WAS NOT INCLUDED in the 1948 Universal Declaration of Human Rights because, at the time, no one could conceive of a world lacking in clean water. It was not to be many decades, however, before the fallacy of this thinking became apparent. Believing that water was indestructible and infinite, people took it for granted and wantonly polluted, mismanaged, and displaced it for our convenience. We used it to grow crops in deserts and to dump waste into oceans and we shipped it out of watersheds in the form of embedded or "virtual" water exports to support a growing global market economy.

The fight to have the United Nations recognize that there should be a human right to water took at least two decades and involved many dedicated people and organizations. The call came out of the struggles of people in thousands of communities around the world who sought the simple dignity of clean water for daily living and basic sanitation services. They also needed to protect their local water sources from government and corporate abuse.

MDGs — STOPGAP MEASURES

Serious attempts have been made at the United Nations to deal with this crisis, but they are not sufficient. The UN General Assembly adopted a set of Millennium Development Goals (MDGs) in 2000 as part of a commitment to deal with the most egregious aspects of persistent poverty. The commitment on water and sanitation is to halve the proportion of people living without sustainable access to safe drinking water and basic sanitation by 2015. The UN itself admits that it is way behind in achieving these goals for sanitation. Advances in sanitation are bypassing poor and rural communities, it reports, noting that improvements disproportionately benefit the better-off, while access for the poorest 40 percent of households is hardly increasing. At the current rate of progress, it will take till mid-century to provide three-quarters of the global population with improved sanitation.[1]

But the UN claims that it is closer to attaining its goals on drinking water access. The World Health Organization reports that since 1990, 1.3 billion people have gained access to improved drinking water, and that the UN is "on target" to meet or exceed its drinking water target. Many question this assertion. One of the chief measurements of access to drinking water used by the UN is the number of pipes installed in a country. But just because there is a pipe does not mean clean water is coming out of it, and even if there is, it may be far from where people actually live. Further, if tariffs on the water are too high and cannot be paid, new pipes are immaterial. I have personally witnessed people turn away from brand-new pipes carrying clean water because access required money for prepaid water meters; they headed instead to rivers with cholera warning signs along their banks.

As well, even as governments move to meet these targets, the declining global water stocks are bringing new communities into crisis. Professor Asit Biswas, president of the Third World Centre for Water Management, calls this claim of success "baloney" and predicts that by the UN deadline of 2015, more people in the world will be suffering from the water crisis than when the goals were first adopted.[2] Says Catarina de Albuquerque, a special rapporteur advising the Human Rights Council on the human right to safe drinking water and sanitation, "I have witnessed the unintended but perverse effect that MDGs can have, making governments feel proud about their achievements regarding the MDGs, while unfortunately forgetting about the poor, migrants, slum dwellers and ethnic minorities who still lack access."[3]

Further, these assertions of success fly in the face of other UN reports that suggest the crisis is deepening. For instance, UN-HABITAT reports that by 2030, more than half the populations of huge urban centres will be slum dwellers, with no access to water or sanitation services whatsoever. And a comprehensive report on Africa shows that water availability per person in Africa is steadily declining and that only twenty-six of the continent's fifty-three countries are currently on track to meet the MDG drinking water targets.[4]

POWERFUL ADVERSARIES

While it might seem to be a given that water is a human right, for years many powerful forces have come together to prevent it from being officially recognized. One powerful opponent is the World Water Council, an international water policy think-tank, the bulk of whose 300-plus members are water and engineering corporations, water industry associations, and investment banks. Past president Loïc Fauchon was also past president of Groupe des Eaux de Marseille, owned by Suez and Veolia, the two biggest water utilities in the world.

Every three years the World Water Council holds a large and influential gathering of water experts, private interests, and government officials to set directions for global water policy and financing. Known as the World Water Forum, it has now overtaken any gathering of the United Nations as the pre-eminent global water symposium. Government policy makers and World Bank and United Nations officials pay it great heed. At every meeting since its inception in 1997, the World Water Forum has directly refused to recognize the right to water in the ministerial declaration that is released on the final day.

At the Forum held in March 2009 in Istanbul, Turkey, and attended by 25,000 delegates from 150 countries, leaders again refused to include the right to water in the official ministerial declaration, resulting in a strongly worded rebuke from Miguel d'Escoto Brockmann, then president of the UN General Assembly. Even the ministerial statement that came out of the World Water Forum held in Marseille in March 2012 — almost two years after the UN recognized the human rights to water and sanitation — once again failed to clearly endorse and repeat the resolution, instead using wording that would allow countries to dodge their legal obligation to uphold these rights.

At the heart of the debate was the distinction between water being a *need* or a *right*. This is not simply a semantic distinction. One cannot trade or sell a human right or deny it to someone on the basis of inability to pay. The World Water Council and the World Bank promote private, for-profit water delivery systems, thus encouraging the concept of water as a need that can be filled by private as well as public operators. The *right* to water, however, denotes that water is a basic right, regardless of ability to pay, and boosts the arguments that it should be delivered as a public service.

Major opponents to the declaration included some First World governments opposed to extending new rights such as the rights to water and sanitation, and worried about the cost and accountability involved. In explaining why the United States delegation did not support the right to water, State Department spokesman Andy Laine said, "Establishing an international right to anything raises a number of complicated issues regarding the nature of that right, how that right would be enforced, and which parties would bear responsibility for ensuring these rights are met."[5]

The U.S. and Canada, two historical opponents of the right to water, have recent histories of refusing to recognize what are called "second- and third-generation" human rights. They support "first-generation" human rights such as freedom of speech, the right to a fair trial, freedom of religion, and voting rights — often referred to as "negative rights" and all guaranteed in the 1948 Universal Declaration of Human Rights. But they are less likely to promote the more proactive second generation of rights, such as the rights to employment, housing, health care, and social security. These are often called "positive rights"; some are found in the Universal Declaration of Human Rights but were more fully covered in the International Covenant on Economic, Social and Cultural Rights.

Canada and the U.S. are even less supportive of third-generation rights such as the rights to self-determination and economic and social development, group and collective rights, and the right to protect local natural resources. For these countries, the rights to water and particularly sanitation are political goals disguised as human rights that create an attendant (and unwanted) set of responsibilities.

It is important to note that the countries that most strongly opposed the General Assembly resolution — Canada, the United States, Australia, New Zealand, and the United Kingdom — all favour a market-based economy and have adopted different forms of privatization and commodification of their own water supplies. These countries promote open global trade and investment rights for corporations, a philosophy that has equipped private commercial companies with powerful new tools for asserting their proprietary interests in water and water services.

"Unfortunately, the most significant developments in international law that bear upon the human right to water are not taking place under the auspices of the United Nations," wrote Canadian trade expert Steven Shrybman before the resolutions were adopted, "but rather under the World Trade Organization, and more importantly, under a myriad of foreign investment treaties. Under these regimes, water is regarded as a good, an investment, and a service." As a result, governments are severely constrained from establishing the policies and practices needed to protect human rights, the environment, and other non-commercial societal goals.[6]

A large part of the resistance to the right to water from these countries came from the fact that they support water as a market good in a variety of international, regional, and bilateral trade and investment deals, and they (correctly) perceive a real conflict between the two models.

PERSISTENT FRIENDS

In spite of the resistance of these powerful forces, however, the demand for the recognition of the rights to water and sanitation steadily grew, led by a dynamic international water justice movement and supported by a number of countries from the Global South, particularly South America, and a handful of countries from the Global North. This movement attended every World Water Forum, criticizing the corporate influence behind the meetings and gaining strength inside the forums themselves until it had the capacity to create alternative people's forums that demanded that water be named a common heritage, a public trust, and a human right. The Alternative World Water Forum in Marseille had 5,000 in attendance and rivalled the official forum for media attention.

A key argument of this movement, of which I have been a part, was that lack of access to safe drinking water and sanitation was hampering the realization of a number of other key human rights obligations already adopted by the UN. Slowly the right to water came to be recognized in a number of international resolutions and declarations. The most important of these was General Comment 15, adopted in 2002 as an "authoritative interpretation" of the International Covenant on Economic, Social and Cultural Rights. In it, the right to water is named as a prerequisite for realizing all other human rights, and "indispensable for leading a life in human dignity."

But an interpretation of an existing convention is not the same as a stand-alone instrument. So, in 2006, the newly formed Human Rights Council, led by Spain and Germany, requested that Canadian human rights expert Louise Arbour, then high commissioner for human rights, conduct a detailed study on the scope and content of the relevant human rights obligations and make recommendations for future action. The global water justice movement weighed in. Canada's Anil Naidoo, of the Blue Planet Project, sent a statement to the high commissioner signed by 185 organizations from 48 countries, calling for the appointment of a special rapporteur on water. In the statement they noted that the failure of an existing UN commitment had allowed several nations to deny the inherent right to water of their citizens.

The report of the high commissioner, tabled in October 2007, noted that "specific, dedicated and sustained attention to safe drinking water and sanitation is currently lacking at the international level" and recommended that access to safe drinking water and sanitation be recognized as a human right.[7] In September 2008, the Human Rights Council appointed Catarina de Albuquerque, a high-energy human rights professor from Portugal, to be the first independent expert on human rights obligations relating to access to safe drinking water and sanitation. Her term was renewed in 2011 and her title changed to Special Rapporteur, giving her clearer authority. Having someone named to follow and foster this process was an important step in moving towards formal recognition of these rights.

INSIDE STRATEGY

I have been deeply involved with the global water justice movement, attending every World Water Forum and helping to create the alternative forum process. I have also been immersed in the fight for recognition of the right to water and helped form an international organization called Friends of the Right to Water.

In 2008 I was given the honour of serving as senior advisor on water to Miguel d'Escoto Brockmann, the newly elected sixty-third president of the UN General Assembly. President Brockmann (or Father Miguel, as he prefers to be called) is a Nicaraguan priest by training with a long history of deep commitment to social justice issues. On taking office, he publicly declared his concern about the impact of the global water crisis on the poor and pledged his support for a General Assembly resolution recognizing the human rights to water and sanitation. Without his direction and support, I doubt that we would have been successful.

President d'Escoto Brockmann contacted me just before he took office and invited me to advise him and guide the process towards a General Assembly resolution. It was ironic in a way that I should be chosen for this task, as it was the Canadian government under Prime Minister Stephen Harper that led the opposition to any formal recognition of the right to water at the United Nations. When I spoke at committees and panels, the Canadian representatives generally left the room.

Soon I was immersed in the complicated politics of the United Nations. I met with all the agencies that deal with water, and many of the ambassadors and policy experts as well, and was shocked to discover the lack of coordination and political leadership around the water crisis. It became clear to me that there were many concerns at the UN about water.

One was that, while the General Assembly is really the only body appropriately situated to deal with the global water crisis, the issue had not been specifically included on the Assembly's agenda. As a result, there was a great disconnect between the often excellent work being done by the agencies and the General Assembly itself. The lack of political will and leadership in the Assembly meant that policy recommendations from UN staff were not being translated into action. As a case in point, the 1992 Rio Earth Summit targeted water, climate change, biodiversity, and desertification for action. By the time of my appointment, the United Nations had addressed all the issues but water with a convention and a plan. Water was the only holdout.

Another concern was that the big water transnationals held key positions of influence at the United Nations, and most were opposed to the right to water. The CEO Water Mandate is an initiative of the

UN Global Compact, a UN–corporate partnership aimed at getting corporations to improve their water practices and policies. But many of the corporations involved in the mandate, including Suez, Nestlé, Coca-Cola and PepsiCo, are themselves objects of severe criticism for their exploitation and commodification of water. Others include companies with poor corporate reputations, such as Dow Chemical, manufacturer of napalm and Agent Orange, and Shell Oil, the target of decades of protest for their fouling of the waters of Nigeria. Recently, even the UN's own watchdog, the Joint Inspection Unit (JIU), warned that some large corporations are using the UN brand to benefit their business and expand public–private partnerships while not conforming to UN values and principles. The JIU called on the General Assembly to rein them in.[8]

Environmental researcher Julie Larsen, a former United Nations Association in Canada youth delegate to the UN, now works at the UN and served on the Secretary-General's High-Level Panel on Global Sustainability. In a detailed report on the influence of the private sector on water policy at the UN, Larsen expressed serious concern over the blurring of boundaries between the public and private sectors and urged the UN General Assembly to prioritize water governance and become the central decision-making authority, independent of private-sector influence, for UN policies and programs in this area.[9]

As the water advisor, I identified allies within the UN and pulled together a team to move the agenda forward. Father Miguel was supportive and gave me every opportunity to present my case. On April 22, 2009 — Earth Day — I spoke to the entire UN General Assembly, beside the great Brazilian theologian and philosopher Leonardo Boff. It was one of my proudest moments. I called on the nations of the world to see water as a public trust and to protect it from private ownership. And I called for formal recognition of the right to water in the near future:

The equitable access to water should also be enshrined once and for all in a United Nations covenant and in nation-state constitutions. A United Nations right to water covenant would set the framework for water as a social and cultural asset and would establish the indispensable legal groundwork for a just system of distribution. It would serve as a common, coherent body of rules for all nations and clarify the right to clean, affordable water for all, regardless of income. A UN right-to-water covenant would establish once and for all that no one *anywhere* should be allowed to die or forced to watch a beloved child die from dirty water simply because they are poor.

No country was stronger in its support than Bolivia and its president, Evo Morales. Bolivia was one of twenty countries that challenged the ministerial statement of the 2009 Istanbul World Water Forum because of its refusal to recognize the right to water. Bolivia's ambassador to the UN was Pablo Solón, a passionate human rights advocate and son of the famous Latin American muralist Walter Solón Romero, who depicted his people's suffering in his powerful art. I had worked with Pablo before he was ambassador to the UN, and on one of my trips to Bolivia, he introduced me to the newly elected President Morales. I put my case to him that Bolivia should lead the crusade at the UN for the right to water, and Evo Morales gave me his word that he would.

In June 2010 Ambassador Solón presented to the General Assembly a draft resolution on the human rights to water and sanitation, to intensive criticism. Many member states fought the process, arguing that the world was not ready for such a step. Some counselled waiting for more research. Others tried to get Solón to weaken the clear language of the resolution and drop the reference to sanitation. Solón refused, rightly citing that lack of sanitation is a leading cause of death and must be included. Others demanded that he add the words "access to water and sanitation," but again Solón refused, noting that changing the wording to include access would let states off the hook, allowing them to argue that private companies were offering those services, and therefore their own obligations had been fulfilled.

Pablo Solón wouldn't budge. He declared that he would rather have the resolution defeated than diluted and that he would like to see which countries would stand up in the General Assembly Hall of the United Nations and vote against the human rights to clean water and life-giving sanitation. While the debate raged on, I and my staff (Anil Naidoo in particular, who had more or less moved to New York for the two months before the vote in order to work for its ratification) and our allies were lobbying individual member states to gain the needed votes.

GETTING IT RIGHT

On July 28, nearly two months later and after much wrangling, thirty-nine countries, mostly from the Global South, presented the final resolution to the General Assembly. It said: "The General Assembly recognizes the right to safe and clean drinking water and sanitation as a human right that is essential for the full enjoyment of life and all human rights," and it called upon all member states and international organizations to assist developing countries "to provide safe, clean, accessible and affordable drinking water and sanitation for all." It also invited the special rapporteur to report annually to the General Assembly.

One hundred and twenty-two countries, including China, Russia, Germany, France, Spain, and Brazil, supported the resolution, and many of those that abstained said they would revisit their opposition if the Human Rights Council were to weigh in with a similar resolution. The support the resolution received that day demonstrated that the world was finally moving to address the issue. The countries that voted in favour represent 5.4 billion people; those that abstained represent 1.1 billion.[10]

The countries that opposed the resolution did not actually vote against it; rather, they abstained. This was the best signal yet that the debate over these fundamental rights was finally winding down. Despite this, the abstaining countries made a number of angry speeches after the vote, accusing Bolivia of forcing a vote that no one was ready for. Solón stood with his arms folded across his chest and a small smile on his face, letting the bitterness pass him by like so much air.

On September 30, 2010, the forty-seven members of the UN Human Rights Council adopted a second resolution affirming the human rights to water and sanitation, making it binding upon governments and setting out their obligations and responsibilities. In a welcome and surprise move, the United States — a new member of the Human Rights Council — declared that it was "proud" to take the significant step of joining consensus on this resolution.

The circle was completed at Rio+20, the June 2012 United Nations conference on sustainable development. After a strenuous campaign, the rights to water and sanitation were included in *The Future We Want*, the official statement of the summit. Even Canada, the last holdout, signed the document, signalling the end of the debate.

The two resolutions, together with the clear statement of recognition at Rio+20, represented an extraordinary breakthrough in the international struggle for the rights to safe, clean drinking water and sanitation and a crucial milestone in the fight for water justice. Now the hard work to make it real was to begin.

3

IMPLEMENTING THE RIGHT TO WATER

The right to water and sanitation is a human right, equal to
all other human rights, which implies that it is justiciable and
enforceable. Hence from today onwards we have an even greater
responsibility to concentrate all our efforts in the implementation and full realization of this essential right. —**Catarina de
Albuquerque, Special Rapporteur**

T HE HUMAN RIGHTS COUNCIL'S confirmation of the rights to
water and sanitation was an important follow-up to the General
Assembly resolution. And the Council went further than the General
Assembly, specifying that the rights to safe drinking water and sanitation are part of international law. It found that these rights are tied
to the right to an adequate standard of living and related to the right
to the highest attainable standard of physical and mental health and
the right to life and human dignity. Quite simply, as de Albuquerque
says, recognition of the rights to water and sanitation requires that
these services be available, accessible, safe, acceptable. and affordable
to all, without discrimination.

The Centre on Housing Rights and Evictions (COHRE) has for

many years been a leader in the struggle at the UN for the rights to water and sanitation. It argues that these rights are key to the excluded, because they give priority to people who do not have access to water and place the onus on governments to ensure water services for all. In too many cases, governments construct expensive services that supply a small, privileged fraction of the population rather than low-cost alternatives that would provide water for the majority.

Access to clean water is now a legal entitlement rather than a charity or a commodity, and individuals and groups can hold their governments to account. The right to water prevents the deliberate discrimination against and neglect of vulnerable and marginalized communities by governments or local authorities that might other-wise act to exclude such communities that are seen as undesirable. Impoverished communities can take a larger role in decision mak-ing because the ramification of the resolution is that governments must consult with the communities affected by water service deliv-ery and the conservation of local water resources. Governments and the international community are now accountable, and UN human rights institutions can monitor the implementation of their commit-ments and point out publicly when they have failed.

On the right to sanitation, the definition is clear. The special rapporteur has defined sanitation as "a system for the collection, transport, treatment and disposal or reuse of human excreta and associated hygiene."[1] To meet human rights requirements, sanita-tion must effectively prevent human, animal, and insect contact with excreta, and toilets and latrines must provide privacy and a safe and dignified environment for all. They must be physically accessible, within reach or in the immediate vicinity of each household, educa-tional institution, or workplace, and available for use at all times of the day or night, along with associated services such as the removal of waste water and latrine exhaustion.

As well, they must be affordable, not reducing the individual's or household's capacity to acquire other essential goods and services such as food, education, and health care. Finally, they must be culturally sensitive, using appropriate local technology and giving attention to gender sensitivity and the need for separate male and female public facilities.

Water justice activists have been concerned that neither the General Assembly nor the Human Rights Council excluded governments from contracting out water services to the private sector. But Ashfaq Khalfan, human rights lawyer and economic, social, and cultural rights policy co-ordinator for Amnesty International, notes that the provisions of the Human Rights Council resolution call for full transparency and the free and meaningful participation of local communities. Human rights must be factored into all impact assessments throughout the process of ensuring service provision.[2]

Amnesty International supports effective regulations for all service providers, in keeping with the human rights obligations of member states. Delegation of safe drinking water and sanitation to a third party such as a private utility does not exempt the state from these obligations. In other words, privatization of water services may not be banned, but it now comes under a scrutiny unlike anything it has experienced in the past.

OBLIGATIONS ON GOVERNMENTS

Whether or not they voted for the two resolutions, every member nation of the United Nations is now obligated to accept and recognize the human rights to water and sanitation. As a result, every member nation must take the necessary steps as soon as possible to ensure that everyone in the country has access to water and sanitation, recognizing that some governments will need more time and assistance than others to meet these goals.

While no country is required to share its water resources with another, there is an understood obligation that wealthier countries will contribute the international assistance necessary to complement national efforts in developing countries. And every member nation is required to prepare a "National Plan of Action for the Realization of the Right to Water and Sanitation" and to report to the UN Committee on Economic, Social and Cultural Rights (CESCR) on its performance in this area.

Each member country is expected to develop appropriate tools and mechanisms — which may include legislation — and comprehensive plans and strategies to fulfil the new obligations, particularly in areas where water and sanitation services don't exist at all. The planning and implementation process must be transparent and open to the participation of local communities.

Three obligations are imposed on states with the recognition of a human right:

1. The first is the obligation to respect. Every government must
    refrain from any action or policy that interferes with the rights
    to water and sanitation. This means that no one should be
    denied essential water services because of an inability to pay. In
    communities where water has been privatized, for instance, and
    the price of water has risen out of reach of the local population,
    water provision cannot be removed. In terms of sanitation, the
    obligation to respect means that governments must not prevent
    people from accessing sanitation by arbitrarily interfering with
    customary or traditional arrangements for sanitation without
    providing acceptable alternatives.

2. The second is the obligation to protect. Every government is
    obliged to prevent third parties from interfering with enjoyment
    of the human right to water. Governments now have the obligation to protect local communities from pollution and inequitable

extraction of water by corporations or governments. Citizens and communities can now begin to hold their governments account-able if mining, energy, or fracking companies are destroying their local water sources. In terms of the right to sanitation, gov-ernments are obliged to ensure that private individuals or groups do not prevent anyone from accessing safe sanitation, for exam-ple, by charging excessively for the use of toilets.

3. The third is the obligation to fulfil. Every government is required to adopt any additional measures directed towards the realiza-tion of the right to water. This means that governments must facilitate access by providing water services in communities where none exist, and they must ensure that appropriate stan-dards and regulations are in place to assist individuals with constructing and maintaining toilets. Where individuals or groups are unable to provide water and sanitation services for themselves, governments must provide the necessary assistance, including information training and access to land.[3]

HOLDING GOVERNMENTS ACCOUNTABLE

Governments must now recognize the right to water and sanitation in their own constitutions or laws; the rights will not be completely in place until they are recognized in domestic legislation and consti-tutions. Some countries have already amended their constitutions. South Africa included water as a human right in its new constitution when Nelson Mandela formed his ANC government, and others such as Ethiopia, Ecuador, Kenya, Bolivia, and the Dominican Republic followed. Municipalities in South Africa are required to provide a minimum of twenty-five litres of water a day to individuals or six thousand litres a month to households.

In 2004, after a successful referendum, Uruguay became the first country in the world to vote for the right to water. The language of the constitutional amendment that followed not only guaranteed water as a human right but also said that social considerations must take precedence over economic ones when the government makes water policy. It also affirmed that water is a public service to be delivered by a state agency on a not-for-profit basis. In early 2012, Mexico announced it would amend its constitution to recognize the right to water — a huge breakthrough that came after an intense campaign led by the Coalition of Mexican Organizations for the Right to Water.

Other countries, such as the Netherlands, Belgium, the United Kingdom, and France, have adopted state resolutions that recognize the right to water of their people. To celebrate World Water Day 2012, El Salvador adopted a new law recognizing the right to water, again in response to a citizen-led campaign. To meet its UN commitments, Rwanda's government pledged to provide its entire population with water and sanitation services by 2013, and it is well on its way to achieving that goal. Even some sub-national governments such as California have introduced right-to-water laws, and some regional blocs are moving too.

In January 2011 the European Parliament's Subcommittee on Human Rights held a hearing on compliance of member states with the new rights. On World Water Day 2011, it issued a statement reaffirming its support for the rights to water and sanitation as "part of the human right to an adequate standard of living." Even the Vatican recently recognized the human right to water, adding that water is "not a commercial product but rather a common good that belongs to everyone."[4]

As important as these examples are, they represent just the beginning of the awareness needed to tackle this crisis. Many countries have not moved at all to meet their new obligations. Chile, for instance, voted in favour of the General Assembly resolution but continues to privatize its water, to the advantage of foreign mining companies over local farmers and indigenous people. In spite of its constitutional recognition of the right to water, South Africa permits prepaid water meters and cut-offs for those unable to pay. On behalf of local residents, the advocacy group AfriForum is suing several municipalities in Limpopo province for failing to deliver even the most basic water services.

Spain, although a strong supporter of the right to water at the UN, promotes the growth of water-intensive tourist resorts and golf courses. Municipalities are being told they need to raise water tariffs 100 percent in order to pay for the expense of importing and desalinating water for their citizens, and water is being cut off for residents who cannot pay their water bills. At the same time, however, reports show that tourists have almost four times the daily water consumption of the average Spanish city dweller.[5]

Domestic water justice groups are not waiting for tardy or hostile governments to take the next steps. They are writing up plans to implement the rights to water and sanitation in their countries and are using these plans to pressure their governments. In some countries, including Ecuador and Argentina, bad mining practices have undermined initiatives to protect the environment and strategic sources of water, local water justice groups have reported.

The Indonesian constitution of 1945 stated that water and land "are to be controlled by the state to be used to the greatest benefit of the people," and in 2005 a court interpreted this and several other constitutional provisions as granting recognition of the right to water. Indonesia also voted in favour of the UN General Assembly resolution. But KRuHA, the Indonesian People's Coalition for the Rights to Water, says in a report for the Blue Planet Project that policies implemented by the government promote commercialization of water services. During times of conflict between local communities and corporations, the government sides with the latter.

The report highlights a struggle in Banten, where a Danone bottled-water plant is draining local water sources. The amount of water extracted at the company's new plant will reach more than 5 million litres a day, giving the company a daily profit of almost $2 million. Jakarta's infamous water privatization, negotiated under former dictator Suharto, gave water giants Thames Water and Suez the contract to run its water system; the people got "fourteen years of the most expensive dirty water in the world." Is the government the guardian or the enemy of the people when it comes to the right to water, the group demands to know, and adds that the true test of the government's commitment will be in how it deals with these companies.[6]

Privatization of water services in India violates the human right to water, says Justice Rajinder Sachar, former chief justice of the Delhi High Court. "There is nothing above the Constitution," he stated at a March 2013 conference on Delhi's intention to privatize its water services. "The Preamble says India is a secular, socialist, Republic...and handing over ownership of water to private companies is cheating the Constitution."[7] Yet the country is rapidly moving to privatize its water services. The Indian National Alliance of People's Movements reports that persistent pressure from international financial institutions to deregulate and privatize water services works against India's stated commitment to the human right to water.

In Europe, one million people still do not have access to water and sanitation, yet the campaign to close this gap has been met with a countermove to sell off public assets. A coalition of European water justice organizations reports that powerful fiscal austerity proponents in Europe have used the economic crisis there to impose "cost-saving" measures, including the privatization of water services and even the sale of some state water companies, resulting in dramatic increases in water tariffs.[8]

The same struggle in the United States is set against a backdrop of increasing poverty, deepening inequality, and rising water rates. Roughly one in four children live below the poverty line, and families across the country are scrambling to meet basic needs. When she visited the U.S. on a 2011 fact-finding mission, UN special rapporteur Catarina de Albuquerque described a visit to a homeless community that was deeply upsetting. She met "Tim," who called himself the "sanitation technician" for the community. Tim collects bags of human waste, varying in weight from 130 to 230 pounds, and hauls them by bike to a local public restroom a few kilometres away, where he empties the contents in a public toilet, disposes of the plastic bags in the garbage, and sanitizes his hands with water and lemon. "The fact that Tim is left to do this is unacceptable," said de Albuquerque, "an affront to human dignity and a violation of human rights and it must be stopped."[9]

U.S.-based Food and Water Watch published a report that identified rural Americans, Latinos, Native Americans, and African Americans as especially vulnerable segments of the U.S. population that lives without secure access to clean drinking water and functional sanitation systems. "These segments of the population experience a disproportionate lack of access to water and sanitation," said Executive Director Wenonah Hauter. "We can't take our access to safe and affordable drinking water and sanitation for granted — even here in the United States."[10] Of the tens of thousands of water cut-offs in Detroit, the vast majority are to people of colour. The Boston Water and Sewer Commission reports that for every 1 percent increase in one Boston ward's population of people of colour, the number of threatened cut-offs increases by 4 percent.

The Council of Canadians reports that the rights to water and sanitation are routinely violated in First Nations communities in Canada, and that First Nations homes are 90 percent more likely to be without running water than the homes of other Canadians. The country was galvanized in the winter of 2013 when the chief of an impoverished First Nations community staged a hunger strike, demanding action from the Government of Canada. Chief Theresa Spence, of the Attawapiskat First Nation in northern Ontario, took up residence in a tepee at an outdoor indigenous education centre in Ottawa and refused solid food for six weeks. Her community lives in extreme poverty, where drinking water supplies are untreated and unusable. Many residents have no sanitation in their homes either. Chief Spence's hunger strike helped inspire a nation-wide First Nations youth movement called Idle No More, which drew media attention from around the world. The Council of Canadians and the Assembly of First Nations called on the Canadian government to honour its UN commitment and provide safe, accessible drinking water to its First Nations peoples.

All of the reports from which I have drawn these disturbing examples have been shared with the UN special rapporteur in order to further her work and to spur the United Nations to take more action.

USING THE COURTS

Water struggles underwrite the conflicts in the Middle East. The Negev Bedouin are a nomadic people who lived in the deserts of what is now southern Israel and were forced from their traditional lands as permanent cities and farms developed around them. As Israel grew, many of them were settled in townships, some without access to electricity or running water. In 2010, several Bedouins living in "non-recognized" communities in the Negev Desert took their fight for the right to water to the Supreme Court of Israel. Seeking to overturn a 2006 ruling that rejected their application to be connected to a Mekorot water company main, six Bedouin residents argued that their lack of access to water violated their basic human rights. In a

June 2011 ruling, the Supreme Court agreed, saying that water is a "basic human right deserving of constitutional protection by virtue of the constitutional right to human dignity" — language straight out of the Human Rights Council resolution.[11]

For several decades the government of Botswana had been forcibly and violently evicting Kalahari Bushmen from their traditional lands in the Central Kalahari Game Reserve. But the Bushmen kept returning to the desert, unable to live well outside their ancestral homeland. In 2002, in a particularly vicious move, the government smashed their only major water borehole to secure the eviction of those who remained. Diamonds had been discovered on their lands, and the government handed over priority land and water access to the mining companies and ecotourism operators. In his book *Heart of Dryness*, American journalist James Workman wrote about his time among the Kalahari Bushmen and the suffering they endured, living in a desert without water and in constant fear of the security forces that were terrorizing them.

Workman recounted the haunting story of Qoroxloo, a Bushman elder, and her small band. The authorities arrived in trucks carrying big guns to destroy their remaining water canteens and spill their water barrels. Despite this, the people would still not leave their land. As the thirsty months passed, water had to be scrounged from dew and wild melons. Anyone caught trying to smuggle water into the desert was arrested and imprisoned. Hunting was outlawed as well. The plan was to deprive the people of any form of water or sustenance so that they would leave. But Qoroxloo stayed and looked after her band, giving all the water she could to the young. One hot day in 2005, her family found her dead under a tree. An autopsy revealed that Qoroxloo had almost no fluid in her body and that her heart was completely dried up. She had sacrificed her own water needs for years so that others might live, and she died of dehydration.[12]

These were the stakes when the Bushmen, working with Survival

International, took their government to court. In 2006 they won an important victory that allowed them to return to their ancestral homeland. But in that decision they did not win back the right to their water sources, so the Bushmen appealed to gain access to their broken borehole. One week before the UN General Assembly voted to recognize the right to water and sanitation, a High Court judgement again denied the Bushmen their water rights. They did not give up. Then a momentous decision came down from the Court of Appeal in January 2011, citing the UN's new recognition of the rights to water and sanitation. The court unanimously quashed the earlier ruling and found that the Bushmen did have the right to use their old borehole, as well as the right to sink new boreholes. The ruling judges called the treatment of the Bushmen by the government "degrading."

In its judgement, the court said it was "entitled to have regard to international consensus on the importance of access to water" and referenced the two UN resolutions. Roy Sesana, a Bushman leader, received the 2005 Right Livelihood Award (known as the "alternative Nobel") for his fight for justice for his people, the same year that Tony Clarke and I received it for our work on water justice. At a formal ceremony held in the stately six-hundred-year-old Parliament Buildings in Stockholm, lit by torches and candles on a cold December night, Sesana accepted his award while dressed in traditional loincloth and headdress, holding a spear. On the day the borehole was reopened, he stood beside it along with his community and said, "We are very happy that our rights have finally been recognized. We have been waiting a long time for this. Like any human beings, we need water to live."[13]

CRUCIAL NEXT STEPS

While there have been good-news stories since that of the Bushmen, much work remains to be done. The world community must come together to bring justice and equality to the issue of access to water in a time of rising demand. To do this, we need to build on the work that has gone before and expand the scope of the obligations recognized by the General Assembly and the Human Rights Council. Many governments will make the narrowest interpretation possible of their obligations; it is imperative that there be a countervailing force against the growth of privatization and commodification.

Using the obligation to respect, we need to assert that no government has the right to remove existing services, as the government of Botswana did to the Kalahari Bushmen; as authorities in Detroit, Michigan, did to tens of thousands of residents, cutting off their water supply when rising water rates made it hard for them to pay their bills; or as the City of Johannesburg does when it denies water to residents unable to pay for water meters.

Using the obligation to protect, we need to challenge any laws or practices that remove or contaminate local water sources. These encompass the public auctioning of water rights in Chile to foreign companies, leaving local farmers and indigenous peoples without water, and sand mining in Tamil Nadu, India, where sand removed from local rivers for urban construction is destroying watersheds. They also include fracking in New York State, where local watersheds are at risk of severe contamination, and dam construction in Turkey, where rural communities and their land and water are being submerged. These and many other actions violate the rights of local people to uncontaminated water sources.

Using the obligation to fulfil, we need to demand extension of public water and sanitation services to those communities and people not now served, regardless of their ability to pay. This will mean reprioritization of domestic and international economic and development policies. In many communities of the Global South, for instance, tourists have far more access to clean water and sanitation than do local residents. Even in wealthier areas such as the Mediterranean, tourists use water needed by the local population. The human right to water can be used to challenge these practices that favour certain groups over others.

Huge new sources of groundwater have been documented in Africa, and the fight is on for control of this water. If these sources are not harnessed for the good of all the people and communities of Africa and are allowed instead to become the property of transnational corporations, daily life may not change for the vast majority of Africans, who will still have little access to affordable water. The right-to-water principle — as well as the principle that this water is a common heritage, not private property — must determine the future of these water supplies.

Some governments favour military budgets and industrial and resource exploitation over providing basic services to their people. Annual global military spending now stands at $1.74 trillion, having increased for thirteen years in a row.[14] Yet the UN estimates it would cost only about $10 billion to $30 billion a year to meet minimum standards of the Millennium Development Goals on water and sanitation, providing half of those who need it with these services. According to the United Nations Development Programme, this money represents less than five days of global military spending, and less than half of what rich countries spend per year on bottled water.[15]

Governments also often give large grants, tax breaks, and subsidies to extractive industries that destroy local water sources. They make money through the exploitation of their natural resources, money that should be used to implement their legal responsibility to provide clean water and sanitation to their people.

Nnimmo Bassey is a Nigerian environmental rights activist and poet, chair of Friends of the Earth International, and one of *Time* magazine's Heroes of the Environment in 2009. Bassey points out that his government gets huge revenues from oil production. But, rather than use those revenues to provide its people with water, it turns to international aid agencies and the World Bank to support water provision. They in turn award the contracts to private-sector corporations. The World Bank should stop this vicious cycle and use its financial levers more productively. In countries such as Nigeria, it should try to rein in corruption and require that resource revenues be used to promote public services.[16]

We need to exert pressure for the right to water and sanitation everywhere we can, both in other UN venues and conferences and at the International Criminal Court, where we could argue that withholding water access to civilians during a conflict is a war crime, and encourage citation of the resolution in other resolutions and treaties, making it a living concept. Additionally, the Committee on Economic, Social and Cultural Rights, the body that monitors implementation of the international covenant, has recently been authorized to consider individual human-rights communications relating to the covenant, including concerns about denial of the rights to water and sanitation.

PUTTING THE MOST VULNERABLE AT THE CENTRE

As we move forward, it is important to keep in mind that women are disproportionately responsible for water management within their families and communities, and differently affected by the absence of clean water and lack of private sanitation facilities. Yet the United Nations says that women are continuously left out of the policy- and decision-making spheres.[17] Women must be brought into the decision making at every step, and policies must address the different needs of women and girls, including the additional threats of domestic violence attached to water scarcity.

Workers are another important piece of the puzzle. When water services are privatized, public-sector unions lose members as workers are laid off in the private company's hunt for profit. Public Services International, the global federation of public-sector unions, has been tireless in its fight for the rights of workers and their families and the need for governments to provide clean, public, accessible water for all. It is crucial that unions and their members be active participants in the struggle for the rights to water and sanitation, and for other groups to support their rights to good working conditions and fair remuneration.

As big urban centres expand and look for new sources of water, they claim — sometimes violently — the water supplies of rural communities. Corporations also take advantage of smaller populations in rural areas to claim or pollute local water sources. An inseparable element of the rights to water and sanitation is control and sovereignty of local communities over their water sources and watersheds.

Indigenous peoples, peasants, and subsistence farmers are frequently victims of water theft, water contamination in their territories, and forcible displacement from their lands and watersheds. To address this inequality, we need to explore ways to broaden the range of rights related to water and sanitation to include third-generation rights such as the right to self-determination, group and collective rights, and the right to local natural resources. This would recognize the legitimate concern of many traditional cultural communities that the UN system is tied to a Western notion of individual rights at the expense of other, more collective approaches to advancing human rights.

The United Nations Declaration on the Rights of Indigenous Peoples is an excellent example of third-generation rights in that it includes among its stated rights self-determination; political, legal, economic, social, cultural, and spiritual institutions; traditional knowledge; dignity and well-being; conservation and protection of natural resources on indigenous territory; and free and informed consent for any resource project affecting them. Human rights are not static; they adapt as our understanding of justice grows. Water and sanitation offer us an opportunity to explore this notion of rights that one day could extend to water itself.

4

PAYING FOR WATER FOR ALL

The choice is clear—do the politicians only want to support banks—or will people's lives be just as important? This is an ethical question—and a political one. Water and sanitation must have the highest priority during the crisis.—**Public Services International**

CRITICS OF THE RIGHT to water claim that it gives the green light to waste water. Calling water a human right, they say, allows anyone to use all the water they want for any purpose and will lead to exploitation. Further, they say that those promoting the right think that water should be "free," and anything that is free will not be cared for. This is simply not true. The United Nations is clear that the human right to water is intended to guarantee water for personal and domestic use. There is no "right to water" for the purpose of a green lawn or a swimming pool or a golf course. In fact, to ensure water justice in a world of growing demand requires fierce protection of the world's dwindling water supplies and strict, open, and fair rules around access.

FUNDING DEVELOPMENT AID

Nevertheless, the question of how to pay for the demand, not just for clean water but also for the infrastructure needed to provide sanitation services to the billions who need it, is a crucial one. In addition to the obvious human and community benefits of clean water, there is a direct economic benefit as well. The World Health Organization (WHO) reports that water and sanitation services reduce health risks, free up time for children to attend school, and improve local environments. Almost 10 percent of the global burden of disease could be prevented by universal provision of water and sanitation, says the WHO. Together with increased productivity resulting from a healthier population, meeting the Millennium Development Goals could provide a benefit-to-cost ratio of 7 to 1.

In his 2008–09 biennial report on the state of the world's water, the Pacific Institute's Peter Gleick points out the need to increase funding for the MDG drinking water and sanitation pledge from its current level of about $14 billion a year to $72 billion. The higher estimate takes into account the need for maintenance and upgrading once systems are built. It is not possible to meet the goals with the current allocation, he says, and points out that along with inadequate UN funding, the Organisation for Economic Co-operation and Development (OECD) reports steadily declining international financial assistance for water and sanitation from the wealthy nations.[1]

Many wealthier governments are not meeting their foreign aid goals and have even cut back in past years, citing the need for austerity. In 1970, wealthy countries pledged 0.7 percent of their national income to foreign aid and development. Most never met that target; in fact, in 2010, aid as a share of national income was only 0.32 percent. (In contrast, governments spend 2.6 percent of the world's GDP on the military every year.) Large cuts in 2011–12 aid budgets put hundreds of thousands of the world's poor in danger, says Oxfam

International. Figures from the OECD show that aid from rich countries that year fell by $3.4 billion. "Cutting aid is no way to balance the books," says Jeremy Hobbs, executive director of Oxfam. "Even small cuts in aid cost lives as people are denied life-saving medicines and clean water."[2]

Increasingly, wealthy countries are linking their foreign aid to the well-being of their private sector. The Canadian government under Prime Minister Stephen Harper has tied aid to Canada's global mining companies, financing only those aid and development agencies that will work to promote Canada's mining industry in countries receiving aid. And France has worked directly with developing countries to promote the interests of its water companies, Suez and Veolia, as part of its aid program. This trend is entirely in keeping with a model for financing water and sanitation that puts the onus on the poor to pay, say David Hall and Emanuele Lobina of the Public Services International Research Unit (PSIRU). The orthodox water-aid model sees the state in developing countries as being unable to finance investment and holds out no hope that it will ever be different.[3]

There is latent discrimination in this view. European and North American countries have done very well by a public model of universal water services but often don't promote similar models in the Global South. I well remember a conversation with Belgian and German advisors to the World Bank who were defending public water in their own countries — "because we know how to deliver it" — while insisting that European water companies under the supervision of the World Bank presented the only effective model of water provision for developing countries.

The orthodox water-aid model promotes direct financing of private water companies, with consumers having to foot the bill for increased costs. This is very different from a public service model, in which water and sanitation are seen as essential services delivered by government and paid for by taxes or taxes in combination with service charges and aid funds. The orthodox model undermines the universal human right to water, as those who cannot pay must either do without or have to depend on limited and often politically determined aid funds.

In spite of the fact that every country has now endorsed the human rights to water and sanitation, the World Bank continues to finance foreign private water enterprises instead of funding public services in poor countries. A quarter of all World Bank water funding goes directly to corporations and the private sector, bypassing governments, says a 2012 report by Corporate Accountability International, a Boston-based non-profit that monitors corporate practices around the world. The International Finance Corporation (IFC), the World Bank's private-sector arm, has spent $1.4 billion on private water corporations since 1993, and this money increased to $1 billion a year starting in 2013. While the language of this funding is all about "providing water to the poor," the private interests involved are doing very nicely by their investments. The IFC is attracting $14 to $18 of follow-up private investment for every dollar it invests.[4]

Key to the commercial model is the notion of full-cost pricing — that individuals are responsible for paying for water and sanitation services and should be paying the full cost of their delivery. Both the World Bank and the OECD have called for water pricing to pay for water services. "Everyone said water must be somehow valued: whether you call it cost, or price, or cost recover," said Usha Rao-Monari, senior manager of the International Finance Corporation, after a high-level meeting of industry leaders in Paris. "It's not an infinite resource, and anything that's not an infinite resource must be valued."[5]

David Hall and Emanuele Lobina point out, however, that the public system has huge cost advantages over the private because the state, having the security of a flow of taxation, pays lower interest on loans than the private sector and does not have to turn a profit for

investors. They show evidence that the orthodox approach has failed
to generate significant amounts of private investment in developing
countries, especially in Africa and India, where national, state, and
local governments are financing nearly all the water investment.
In fact, they say, the cost of providing water and sanitation is not
unaffordable in the Global South, and most advances in meeting
the Millennium Development Goals have been made by the countries themselves. The framework that directs aid from the wealthy
countries and the World Bank is not working for anyone but the
private sector.

Importantly, a new, "southern" view of aid is emerging to challenge the traditional northern orthodoxy. This view believes that to
fulfil the obligations of the rights to water and sanitation, governments must provide them as a public service. The World Bank and
the other development banks must stop financing private commercial water services and put aid money directly into public agencies
operating on a not-for-profit model. The practice of the World Bank
and its regional counterparts in promoting privatization of water
services in the Global South must be challenged, and the UN must be
urged to support only public water and sanitation services.

FUNDING DOMESTIC WATER SERVICES

The same issues are being debated by municipal, state, and national
governments that are trying to find funds for domestic water services. There is a need for infrastructure construction and upgrading
all over the world as populations grow and become urbanized.
Globally, water infrastructure costs could run in the trillions of dollars. Environmental concerns have also put pressure on governments
to protect and clean up source waters polluted by decades of abuse.
Recycling waste and storm water is another necessary investment.

All these demands come at a time when most national governments are cutting environmental and public service spending. Many are also promoting public–private partnerships and private water delivery to relieve them of the responsibility of providing water and sanitation. Cash-strapped municipal governments are looking for ways to pay for expensive water protection, upgrading, and delivery at the very time that public funds are becoming increasingly scarce. Most are raising water rates for residential and business users, and many are installing water meters to charge for higher consumption. Some are turning to private utilities to deliver water, arguing the need for private investment funds. As well there is growing pressure from a chorus of economists, politicians, and some environmentalists to put a price on water to pay for the true cost of water services.

If the right to water is to be honoured, it is crucial to keep municipal water services in public hands and to maintain their status as a public service. To do this, we have to find funding sources to cover the true cost of water protection and delivery. If we do not find the money needed by local governments, they are likely to turn to the only source of readily available investment money: private utilities.

The best and fairest way to pay for public municipal water services is through a system of progressive taxation. Most developed countries have traditionally used taxation to create public and universal water services, often out of the realization that maintaining public health requires water and sanitation for all. In the United Kingdom, the majority of households pay annual charges based on the value of their property rather than metered consumption of water. A combination of education, improved infrastructure (including replacing leaking pipes), and legislation to promote water-saving technology will save more water than pricing for conservation alone. In fact, a community with good conservation measures and a progressive tax system does not need to charge much for water services at all.

However, it is a reality that we live in an age of austerity, and very few national governments are going to run on a platform of raising taxes for anything, especially to pay for water services that have, until recently, been inexpensive. So municipalities are charging residential and urban water users higher rates and metering homes in order to charge by volume used. While we water justice advocates have deplored the metering model as user fees and have documented the damage such fees have done to the poor of the Global South, it is also a seemingly unstoppable trend.

Two-thirds of OECD member countries meter more than 90 percent of single-family homes, and the practice is growing rapidly in the Global South. If this process is not to lead to the privatization of services and exclusion of the poor, it is imperative that governments set rules now about how to administer service charges in a fair and just way that maintains public control.

MARKET-ORIENTED PRICING

We need to distinguish between "market-oriented pricing" and "service charges" for residential water and sanitation provision, as well as licence fees for commercial access to raw water. The words *pricing* and *commodification* are often used interchangeably. However, charging for water services is not limited to the private sphere, and does not in and of itself lead to commodification. Charges for water services are already applied within the public system as a source of funding beyond taxation. In the public sector, the money collected goes to pay for improved services and source protection, not profit.

In the private sector, water rates are set in order to generate the profits required to pay private investors. Its proponents see pricing as a cornerstone of the private model of water development and a way to let governments off the hook for provision of water services. They believe it is time for government subsidization of water services to end. For them, water is just a good like any other, and citizens are first and foremost consumers. As a market commodity, water would be put on the open market like oil and gas; the market would set the "right" price for water.

The proponents of the private model also believe in "full cost recovery," in which the consumer pays the full cost of water services, including source protection and infrastructure. If the service has been privatized, full cost recovery also includes profit for the company and its shareholders. This differs from a service charge within the public sector, in that service charges are kept at an affordable level through additional government funds. Pricing based on full cost recovery is part of the market-friendly reforms for water distribution advocated by the World Bank and other proponents of water commodification, and it is a key step in "personalizing" responsibility for water in our world.

Market-oriented pricing violates the right to water. With a sharp increase in tariffs based on full cost recovery, water rates will hit the poor disproportionately. Their water use is more likely to be for the essentials — cooking, sanitation, and drinking water — than that of the more affluent with their lawns, gardens, and swimming pools. Untold millions have had their water cut off in the Global South because of their inability to pay for high-priced water. In a recent report, the Quebec-based water advocacy group Coalition Eau Secours! found that 70 percent of household water use in Quebec is for basic needs.[6] Pricing water at rising market values will force these basics out of the range of many.

As U.S. advocacy watchdog Food and Water Watch explains in its report *Priceless: The Market Myth of Water Pricing Reform*, applying market rules of competition simply don't make sense when it comes to water. Cheaper competitive types of water simply don't exist. "If the price of water is too dear, people could not choose to drink another liquid like ammonia or gasoline."[7]

SETTING RULES FOR SERVICE CHARGES

Water is a human right and a public service, but there can be no human right to water if some are allowed to appropriate water for profit while others do without. Water is a public service that must be delivered on a not-for-profit basis by a publicly owned and accountable agency or, in more traditional societies, by agreed-upon methods of water sharing and distribution. All policy dealing with water funding should promote these principles.

If water service charges are to be levied, no one should be denied water because of an inability to pay for it. Service charges should be supplemented by government funds (and aid money, in the case of developing countries) and not allowed to follow a full cost recovery model. It must also be clear that the charges are for the cost of the service, not the water itself, and the money collected is to go towards source-water protection, infrastructure upgrading, and protection of the rights to water and sanitation for all. Finally, the public should participate in the setting of fair and equitable rates.

Essentially, with minor variations, there are three ways to levy a water service charge: (1) a flat rate per household and business; (2) charging by volume use; or (3) a tiered system of charging by volume use, with lower rates for lower consumption and higher rates for higher consumption. Many environmentalists now argue for volumetric rates based on a conservation-oriented service charge, whereby those who use more water are charged more than those who use less.

With flat volumetric charging, we pay for the water we use — the more we use, the more we pay. Conservation-oriented block charging actually sets the bar very low for basic water needs and then charges more per unit of water used for the second tier of use, and more again for the third. In other words, the basic water we use is cheap; when we start using more, we pay more per unit. Advocates say that not only will this promote conservation, as people will see how much water they are using, but it will also keep prices low for essential uses of water and make people who fill their swimming pools and water their lawns pay much higher rates. They argue that it is a question of fairness: why should prolific water users pay the same amount as those who do their best to conserve?

If a conservation-oriented system of service charges is to be used, it must not discriminate against large families, who will use more water for basic needs, or against those who use water to grow their own food. Nor should it serve as an excuse for unlimited water consumption. When needed, governments must still exercise their right and ability to restrict access for non-essential water use.

CHARGING THE BIG USERS

Putting all the emphasis on municipal water consumers, however, misses the point and lets the real water guzzlers off the hook. Most studies, reports, and position papers on water pricing limit their discussion to residential water users, and most metering targets only municipalities. Yet the amount of water consumed by urban residential and business users around the world is small, under 10 percent. And most municipal water use is not consumptive — that is, it is returned to the watershed.

Ninety percent of water is used by the water-reliant natural resource industries — mostly industrial agriculture but also manufacturing, mining, oil and gas, pulp and paper, and electricity generation. Even high rates for residential users will not cover the cost of water use by these commercial interests. Clearly, targeting residential users alone is a misplaced strategy in fighting water abuse. Any strategy that ignores 90 percent of the problem is inherently flawed.

Most countries charge little or nothing for "raw water." That's because large corporate interests are invested in access to cheap water, and few governments appear prepared to take them on. For industry, therefore, water is generally cheap and easy to obtain. Many commercial water users and industrial firms get their water from the same municipal utilities as household users, and some communities attract industrial activities by offering cheap water rates, often lower than those paid by homeowners and local businesses. In fact, many have a kind of reverse block funding, whereby the more water an industry uses, the less it pays. Hotels, golf courses, and the tourist industry are all subsidized by being charged residential rates or lower, even though they consume vast amounts of water and use local water supplies to run their businesses.

It is time to start charging industrial and commercial interests for the water they use. They are making a profit from what is essentially a public trust. This preferential treatment contributes to the redistribution of wealth from the public to the private sphere. Private companies should be returning some of that profit to the public good. Most governments charge royalty fees for access to other resources. The Canadian province of Alberta, for instance, charges royalty fees to the big energy companies for tar sands oil but lets those same companies use (and destroy) huge amounts of water for free. Why do we not treat water in the same way as energy, forests, and minerals?

This is not a recommendation for market-oriented pricing. Commercial water users should be paying a licence or permit fee for access to a common asset. The money collected could be used for watershed protection and restoration, infrastructure investment, and supplementation of municipal water service charges to ensure water for all. Commercial users would also be more inclined to conserve water if they have to pay for it. The Canadian National Round Table on the Environment and the Economy says that raising the price of water for natural resource industries in Canada alone by just five cents per cubic metre would reduce their water intake by 20 percent.[8]

As well, volumetric fees for commercial users might prove to be an incentive for better water-saving practices. A lower block rate might be offered for industries that convert to solar power or that use closed-loop systems that reuse water. Food and Water Watch points to a number of studies that demonstrate significant water savings when pricing is levied on industry. This is because large industries and commercial users have more flexibility to squeeze wasteful water use and practices out of their operations than a household does. Food and Water Watch says that high-volume industrial and commercial water users need water-pricing reform more than households, both to increase fairness to smaller residential water users and to more effectively promote conservation in industry and agriculture.

POTENTIAL PITFALLS

While in some countries electricity generation is the biggest user of water, it restores most of the water to the watershed, unlike industries that remove water permanently from the watershed or pollute it. Agriculture is by far the greatest consumer of water in the world — 70 to 90 percent — as the water used to grow crops leaves the watershed forever. This is called "virtual water": once used, it is lost to the watersheds.

While some farms are still run as family and community operations, increasingly countries are moving to an agribusiness model of food production managed by giant transnational corporations. Six gene giants control most of the world's seed market and five companies dominate the global grain market. In the United States, four companies control over 80 percent of the beef packing industry.

These private corporations make huge profits for their investors from local water supplies and pay little or no money for this water. In fact, governments often subsidize this sector by funding water diversions for irrigation. Further, some operations such as big factory farms get priority rights from governments; in times of groundwater shortages, they have first call on this water.

It is time to start looking at a differential fee system for food production based on the size of the operation and how much production is for local consumption and how much for export. Local, sustainable food production provides a positive counterweight to the globalization of the food trade and conserves local water sources. There could also be lower block rates for good agriculture and water-efficient practices, such as hydroponic market gardens, drip irrigation, and organic crops. As well, volumetric pricing for large food producers would discourage straight-line increases in water consumption and make it harder for large operators to access huge amounts of cheap water. Larger consumers should be paying more than smaller operators; they have greater capacity for research and innovation, and paying for their volume of use will give these corporations the incentive to pursue new techniques.

Terry Boehm is a grain farmer from Saskatchewan, president of the National Farmers Union in Canada, and a fierce defender of family farms and viable rural communities. Boehm says we need a regulatory system for water use in food production that operates on criteria other than the market and that asks about end goals. Is more water needed for domestic or foreign consumption? Is the water resource going to raise farm incomes or corporate profits? Who will benefit the most? What will be the quality of the water after use? Is the water resource better left undisturbed?[9]

Another concern is that some big companies might assume that paying a fee for water access allowed them to buy their way around environmental rules. Paying for water must *not* give natural resource, commercial, and industrial users the right to take as much water as they want or to pollute it. Paying for water does not get around the need to protect it. Using public water for the bottled-water industry or for fracking operations is not a good use of this common resource, and no amount of money will change that fact.

Charging a serious licence fee to industrial and commercial users could help separate out less desirable companies and practices. Licences could and should be denied to any company that pollutes or overuses local water sources. Big business claims that it wants to help solve the global water crisis. Paying for use of this public asset would be a start.

Finally, a spate of free-trade and investment agreements around the world has given foreign corporations the right to claim water they use for production in other countries. Governments must not relinquish public ownership or grant proprietary rights to private companies in return for charging them for water. It would be best for any water-charging scheme to be very explicit about preserving the public trust and clear that the ownership of water will remain in public hands.

SETTING PRINCIPLED PRIORITIES

Honouring the UN recognition of the rights to water and sanitation is hard work and will cost money. We need to get serious about how to fund it and also how to pay for domestic public water services. Internationally, this means getting serious about meeting the UN development aid commitment of 0.7 percent of GDP. It also means ensuring that the financing of water services in the Global South is put into public, not private, for-profit systems, so that every dollar goes to delivering water. At the domestic and local level, this means applying a progressive tax to help subsidize water and pay for the infrastructure that benefits everyone. It also requires removing the onus from residential users, small businesses, and small farmers and putting it where it belongs — on the big commercial users of raw water.

All this must happen against a backdrop of transformation of the global economy to a more sustainable model, and must factor in a cost for the non-renewable resources used in commercial extraction and production. The quest for global water justice must be tied to an overhaul of our economic priorities. It could lead to a more just and sustainable world in all aspects.

WATER IS A COMMON HERITAGE

This principle recognizes that water is a common heritage of humanity and of future generations as well as our own. Because it is a flow source necessary for life and ecosystem health and because there is no substitute for it, water must be regarded as a public trust and preserved as such for all time, in law and in practice. Water is to be preserved forever for public use and governments are required to maintain the water commons for the public's reasonable good. Therefore, water must never be bought, hoarded, traded, or sold as a commodity on the open market. Nor should the private sector determine access to water. Drinking water and wastewater treatment should be managed as a universal not-for-profit public service. No one can "own" water. While there is an economic dimension to water, the private sector must never be allowed to have control over the earth's water supplies, and must abide within the public trust framework in its dealings with the water commons.

5

WATER — COMMONS OR COMMODITY?

There are people who will buy the water when they need it. And the people who have the water to sell it. That's the blood, guts and feathers of the thing. —**American oil tycoon T. Boone Pickens**[1]

TWO COMPETING NARRATIVES ABOUT the earth's freshwater resources are being played out in the twenty-first century. On one side is a powerful clique of decision makers, politicians, international trade and financial institutions, economic advisors and academics, and transnational corporations who view water as a commodity to be bought and sold on the open market, like running shoes. On the other is a global grassroots movement of local communities, the poor, slum dwellers, women, indigenous peoples, peasants, and small farmers who are working with environmentalists, human rights activists, and progressive public water managers and experts in both the Global North and the Global South. They see water as a common heritage and a public trust to be conserved and managed for the public good. Much of the disagreement between these visions rests on the notion of the commons and whether it still applies in today's world.

DEFINING COMMONS AND PUBLIC TRUST

In recent years some very important work has been done to create renewed awareness of the ancient concept of the commons. In
most traditional societies it was assumed that what belonged to one
belonged to all. Many indigenous societies to this day cannot conceive of denying a person or a family basic access to food, air, land,
water, or livelihood. As recently as two decades ago, large parts of
the world still lived off the land, many in isolation from the global
system of competitive trade, and billions still live their everyday lives
outside the global market. For them, community care of their land,
forests, and watersheds is a way of life, as is the equitable sharing of
nature's bounty.

The late America journalist Jonathan Rowe captured the essence
of the concept:

> The commons is the vast realm that lies outside of both the eco-
> nomic market and the institutional state, and that all of us typically
> use without toll or price. The atmosphere and oceans, languages
> and cultures, the stores of human knowledge and wisdom, the
> informal support systems of community, the peace and quiet we
> crave, the genetic building blocks of life — these are all aspects of
> the commons.[2]

Canadian environmentalist Richard Bocking says that the commons are those things to which we have rights simply by being a
member of the human family: "The air we breathe, the fresh water we
drink, the seas, forests, and mountains, the genetic heritage through
which all life is transmitted, the diversity of life itself."[3] *Commons*
is synonymous with community, co-operation, and respect for the
rights and preferences of others, he adds. Some commons, such as
the atmosphere, outer space, and the oceans, may be thought of as

global, while others, such as public spaces, common land, forests, the gene pool, and local medicines, are community commons. "The commons have the quality of always having been there," says Rowe, "one generation after another, available to all."[4]

How to preserve commons assets and equitably share in their benefits will vary with the type of commons. Some, such as wilderness, should be largely off limits. Others, such as the cultural commons, need to be more inclusive. Those with a physical threshold, including fisheries and the atmosphere, need strictly enforceable sustainable use limits.

There are basically three types of commons. The first category includes the water, land, air, forests, and fisheries on which everyone's life depends. The second includes the culture and knowledge that are collective creations of our species. The third is the social commons that guarantees public access to health care, education, and social services, including pensions and welfare. Since adopting the Universal Declaration of Human Rights in 1948, governments are obliged to protect the human rights, cultural diversity, and food and social security of their citizens. Any access to the use of a common heritage carries a corresponding moral obligation to act as its steward on behalf of all, and the accumulation of any right must not infringe on the birthright of others to their equitable share of the common inheritance.

The doctrine of public trust is the vehicle by which the commons is protected. It underpins in law the universal notion of the commons that certain natural resources — particularly air, water, and the oceans — are central to our very existence and considered to be the property of the public, who cannot be denied access. The trust resources must, therefore, be protected for the common good and not appropriated for private gain. Under public trust, governments, as trustees, are obliged to protect these resources and exercise their fiduciary responsibility to sustain them for the long-term use of the entire populace, not just the privileged who might buy inequitable access.

The public trust doctrine was first codified in 529 CE as the Codex Justinianus, named for its creator, Emperor Justinian I, who said, "By the laws of nature, these things are common to all mankind: the air, running water, the sea, and consequently the shores of the sea." This common law has been replicated in many ways and many jurisdictions, including in the Magna Carta. It has been used as a powerful legislative tool in many countries to provide public access to seashores, lakeshores, and fisheries. In a landmark decision from 1926, the Michigan Supreme Court ruled that U.S. states held title to the lands under navigable waters "in trust for the people," and that the "common right" to fish in these waters was protected by "a high, solemn, and perpetual trust...the duty of the state to forever maintain."[5]

Oliver Brandes and Randy Christensen, of the POLIS Water Sustainability Project of the University of Victoria in British Columbia, add that at its core, the public trust is a background principle of property law that serves to strike an appropriate accommodation between the public interest and private development rights, through requiring continuous state supervision of trust resources. Public trust is a recognition, they say, that private rights to use water, for example, are not granted in a completely unencumbered fashion. They must be obtained through an appropriation system administered by government and with implicit restrictions on unduly and irreparably harming the resource and its associated values. This public trust is a safeguard that prevents monopolizing of trust resources and promotes decision making that is accountable to the public.[6]

ENCLOSURE OF THE COMMONS

But the doctrine of public trust is swimming against a tide of increasing private rights. The integrity and health of the commons were challenged when economic globalization and market fundamentalism began to be promoted in the 1970s as the only model of development for the world. The almost universal acceptance over the next decades of the inevitability of economic globalization, by international institutions such as the World Bank as well as governments of developed countries, combined with a technological and transportation revolution that allowed capital to move freely around the world and goods to be produced almost anywhere.

In mere decades, capital went global and First World domestic corporations expanded their operations to other countries to take advantage of cheap labour, weak environmental laws, and local natural resources. For the first time, transnational corporations gained access to the seeds, minerals, timber, and water resources of even the remotest parts of the earth. With time this corporate access became "protected" by regional and bilateral trade and investment agreements, and corporations gained the right to sue governments if their established "right" of access to these formerly common resources was interrupted or curtailed.

Some refer to this process as the second enclosure of the commons. The first, which started in about 1740, took away peasants' rights to farm, graze, and hunt on lands owned by the nobility in England and Wales. Wealthy landowners used their control of state processes to re-appropriate for their private benefit what had become in practice common land, a process often accompanied by force, resistance, and bloodshed. By the nineteenth century, unenclosed commons had become restricted largely to rough pasture in mountainous areas and to non-arable parts of the lowlands in Great Britain.

Enclosure of the commons took place in the Global South as well. Indian physicist and writer Vandana Shiva, who has led the international fight for food sovereignty, points out that privatization of the commons was essential for the Industrial Revolution, in order to provide a steady supply of raw materials to industry. The policies of deforestation and enclosure of the commons were replicated in the colony of India. In 1865 a law was passed lifting protection of the forests as commons, paving the way for commercial exploitation of both land and forests. The ensuing marginalization of Indian peasant communities' rights over their forests, sacred groves, and common grazing lands was the prime cause of impoverishment for millions of people.

The enclosure of water resources came next, says Shiva, through dams, groundwater mining, and privatization schemes. Seed patenting by transnational corporations posed a direct threat to the biodiversity upon which many thousands of communities depended. This concept of private ownership was foreign to most of the world's rural, peasant, and indigenous communities. Traditional societies often shared common resources; they did not view their heritage in terms of property or as goods that have owners and are used for extracting personal economic benefits. Instead, says Shiva, they viewed these commons in terms of the good of the whole community. "For indigenous peoples, heritage is a bundle of relationships rather than a bundle of economic rights."[7]

Common resource knowledge and sharing allowed the creation of a vast pool of agricultural and plant knowledge. To gain access to this store of knowledge, opponents of the commons had to find ways to attack it. A famous paper written in 1968, called "The Tragedy of the Commons," by American biologist Garrett Hardin, gave philosophical and political momentum to the private assault on the commons.[8] Hardin claimed that if no one owned the commons it would soon be plundered, as no one would be responsible for it. He called for abandonment of the commons, citing a case in which animal herders using a common pasture would eventually overgraze it. While those with the most animals would make more money, eventually everyone, rich and poor, would suffer. Hardin's paper became a rallying cry for privatizing common property. Proponents of privatization cite his essay to this day, and Hardin is regularly studied as part of the core curriculum in universities as proof of the "failure of the commons."

However, Hardin was talking about an "open," or unmanaged, commons and he ignored the capacity of common property management systems to provide for sound and sustainable stewardship where such management structures exist and are nurtured. Traditional communities responsible for shared resources care for them in a highly competent way. It is true that the commons have been plundered by societies under both capitalist and socialist systems. The former Soviet Union exploited its water, forest, and mineral systems mercilessly in the name of progress, while contemporary China has done the same. But in all of these cases, the missing piece is a functioning commons management structure. The "tragedy" is not that there is a commons but that it is not sufficiently protected.

(British writer Fred Pearce adds another critique. Hardin was part of the eugenics movement of his time, aimed at "improving" the genetic composition of the human population. He advocated abandoning the "commons in breeding," and later wrote a paper arguing that rich nations can be seen as "lifeboats" sailing in an ocean of poor people trying to get into the boats. The rich had a duty to be selfish and deny entry to the poor even if they drown, said Hardin, who added giving food and medicine to poor countries was the same as giving them access to their lifeboats.)[9]

Governments, mostly but not exclusively from wealthy countries, bought into the allure of unlimited growth and private ownership of water and other commons. Piece by piece, as states dismantled their social security systems and public control of resources, in the halls of government the private values of exclusion, possession, and personal or corporate gain began to replace the commons' values of inclusion, collective ownership, and community assets. Governments started abandoning their responsibility to provide basic social services, instead contracting them out to private companies.

Many areas once thought to be outside the purview of the market became fair game. The race was on to capture and profit from the land, genetic, water, mineral, and forest resources of the commons, thus turning them into commodities, and to use the air, ocean, and freshwater commons as a dumping ground for waste. This had the added benefit to the private sector of passing the problems created by their actions back to the public to either live with or clean up.

American commons activist and writer David Bollier outlines a number of reasons to be concerned about the increasing market exploitation of our commons. The enclosure of the commons needlessly siphons hundreds of billions of dollars away from the public purse every year. Governments use public funds to subsidize private delivery of essential services such as water and sanitation and to boost the bottom line of corporations. These funds could be better used to invest in and protect the water commons. Privatization fosters market concentration and the dominance of large corporations that have the market clout and political influence to obtain public resources on favourable terms.

Enclosure threatens the environment by favouring short-term profits over long-term stewardship. Corporations find it financially desirable to shift health and safety risks to the public and to future generations. Enclosure also imposes new limits on citizen rights and public accountability as private decision making supplants open procedures of democratic participation. Finally, enclosure imposes market values in realms that should be free from commodification, such as community and family life, public institutions, and democratic processes. The market is like a runaway engine, says Bollier, with no governor to tell it when to stop depleting the commons that sustains us all.[10]

ENCLOSURE OF THE WATER COMMONS

There is no better example of a "runaway market engine" than the corporate cartel now being created to own and profit from the planet's supply of fresh water. The private sector saw the coming water crisis before most. While governments continued to believe that the supply of water was inexhaustible and created economic policies based on this supposed plenty, some corporate leaders realized that in a world of declining water supplies, whoever controlled water would be both powerful and wealthy. Corporate interest in the world's dwindling clean water sources has been building for three decades but has dramatically increased in recent years. Transnational corporations view water as a saleable and tradable commodity, not a common asset or public trust, and are set to create a cartel resembling the one that now controls every facet of energy, from exploration and production to distribution.

As I wrote in *Blue Covenant*, "Private, for-profit water companies now provide municipal water services in many parts of the world; put massive amounts of fresh water in bottles for sale; control vast quantities of water used in industrial farming, mining, energy production, computers, cars and other water-intensive industries; own and operate many of the dams, pipelines, nanotechnology, water purification systems and desalination plants governments are looking to for the technological panacea to water shortages; provide infrastructure technologies to replace old municipal water systems; control the virtual trade in water; buy up groundwater rights and whole watersheds in order to own large quantities of water stock;

and, in some countries, buy, sell and trade water, including sewage water, on the open market."

Agribusiness interests purchase and control local water rights to divert water resources from municipal taps to irrigate cash crops and factory livestock production. Energy companies are buying up sugarcane fields and local water rights in South America in order to control the growing biofuel industry. Mining companies buy up water rights in poor countries and employ the water in a leaching process that uses cyanide to separate the gold and other precious metals from the ore. With every passing day, more and more water is being taken out of the commons and claimed by private interests.

In a blistering report, ecological engineer Jo-Shing Yang, of the University of California, says that Wall Street banks and multibillionaires are buying up water all over the world at an unprecedented pace:

> Familiar mega-banks and investing powerhouses such as Goldman Sachs, JP Morgan Chase, Citigroup, UBS, Deutsche Bank, Credit Suisse, Macquarie Bank, Barclays Bank, the Blackstone Group, Allianz, and HSBC Bank, among others, are consolidating their control over water. Wealthy tycoons such as T. Boone Pickens, former President George H. W. Bush and his family, Hong Kong's Li Ka-shing, Philippines' Manuel V. Pangilinan and other Filipino billionaires, and others are also buying thousands of acres of land with aquifers, lakes, water rights, water utilities, and shares in water engineering and technology companies all over the world.[11]

Willem Buiter, chief economist for Citibank, does not mince words in his promotion of water as a market commodity. He writes:

> I expect to see in the near future a massive expansion of investment in the water sector, including the production of fresh, clean water

from other sources (desalination, purification), storage, shipping, and transportation of water. I expect to see pipeline networks that will exceed the capacity of those for oil and gas today.

I see fleets of water tankers (single-hulled!) and storage facilities that will dwarf those we currently have for oil, natural gas and LNG.

I expect to see a globally integrated market for fresh water within 25 to 30 years. Once the spot markets for water are integrated, futures markets and other derivative water-based financial instruments — puts, calls, swaps — both exchange-traded and OTC will follow. There will be different grades and types of fresh water, just the way we have light sweet and heavy sour crude oil today. Water as an asset class will, in my view, become eventually the single most important physical-commodity based asset class, dwarfing oil, copper, agricultural commodities and precious metals.[12]

Most major banks now have water-targeted investment funds. Public pensions around the world are investing in private water projects. Quebec and Canadian public pension funds have acquired shares in England's private water companies, South East Water and Anglian Water, and the Ontario Teachers' Pension Plan is a major stakeholder in Chile's fully privatized water service system. Business schools around the world teach that the only way to protect water is to put it on the open market for sale, like oil and gas, and sell it to the highest bidder.

Privatization in the United States is expected to explode in the next five years, says a new publication, *U.S. Water Industry Outlook*. The country is poised to see a surge in privatization of water and public–private partnerships, says the report, in a quest to capitalize on a declining resource. This process will include water rights and access to increasing water fees that will attract the interest of private equity and hedge funds that can in turn capitalize on takeover of the public good, say the authors.[13]

WATER RIGHTS AND WATER TRADING THREATEN THE WATER COMMONS

Water trading is a recent form of privatization that challenges the public trust doctrine. Water rights give a user guaranteed access to a local water source; they usually emerge from a person's ownership of land bordering a watercourse or from his or her right to the use of a nearby watercourse. Water rights are often inherited but can be acquired if the government converts water licences to water rights, giving the new "owner" more discretion in their use. Traditionally water rights have not entitled the owner to transport water away from the land abutting the waterway.

However, in the past two decades, trading of water rights has become an emerging business in the United States, Chile, Spain, and Australia as landowners, farmers, and speculators buy and sell water rights and transport the actual water over long distances. Water trading has emerged as a dangerous new form of water commodi-fication. (This market model of water trading is not to be confused with traditional water sharing or with indigenous water exchanges in communities in the Middle East, New Mexico, Africa, India, and Latin America.)

The argument put forward by economists, business schools, and some politicians in favour of water trading is that it will promote more efficient water allocation because a market-based price acts as an incentive for users to reallocate resources from low-value to high-value activities. In practice, however, water trading allows big agribusiness, bottled-water companies, and other big private water users to buy up water rights to use themselves or sell on the open market to domestic and foreign investors. Water that was once a commons and a public trust is now separated from the land and watershed and traded between buyers and sellers — a short step away from a full and open water commodities market.

In the United States, water trading is more prevalent in the western states, where the history of water law is based on "prior appropriation," or "first in time, first in right." Water rights were used in the American (and Canadian) West to encourage early settlement and farming. Shiney Varghese, of the Institute for Agriculture and Trade Policy, says that prior appropriation in the west established not only the quantity of water a farmer could claim but also the purpose for which it could be used. This allowed the rights holder to decouple water rights from land rights, and those holders could treat water as a commodity. The separation of water from land allowed those who inherited water rights to sell their surplus to newer users, creating one of the original water markets.

Concerns over conservation and the environment cannot change the nature of these private rights. Even in times of drought, as long as there is water in the water source, the senior rights holders are able to use or sell their full allocation of water. Varghese writes that while water transfers were minimal in the early years, by the late 1970s all viable options for additional water supply in the western states had been exhausted, and water banking, leasing, and trading started in earnest. In the two decades between 1987 and 2009, more than 4,400 water transactions were recorded in the twelve western states.[14]

However, until recently, water trading in the United States took place within districts, mostly among farmers — but no longer. In water-short California, some farmers are letting their fields lie fallow and are selling water as a cash crop to local municipalities. The state is now facing proposals to allow farmers to sell their water to developers, piping it long distances from its watershed. Two farmers in the San Joaquin Valley are proposing to sell their water rights to a developer for $11.7 million. In 2009 the Dudley Ridge Water District sold $73 million worth of water to the Mojave Water District. Aside from the obvious concern about letting good farmland be taken out of production, the government has already subsidized most of these farmers for the water they now want to sell for profit.

In 2011, Texan billionaire oilman T. Boone Pickens sold 16 trillion litres of water rights he had bought from the Ogallala Aquifer to a Texas water supplier for $103 million. His company, Mesa Water, has come under strong criticism for hoarding water and selling it for profit in an area desperate for water. Pickens compared the deal to "buying and selling a motor boat."[15]

In Spain, developers and the tourist industry are dramatically increasing the demand for water. At the same time, encouraged by the government to engage in water trading, farmers have switched to bigger, more industrial operations and water-intensive crops that are highly unsuited to a semi-arid region. As whole areas of the south predictably started running out of water, farmers started fighting developers over water rights. Farmers are buying and selling water "like gold" on a rapidly growing black market, mostly from illegal wells, reports the *New York Times*. The hundreds of thousands of wells — most of them illegal — that have in the past provided a temporary reprieve from thirst have depleted groundwater to the point of no return, says journalist Elizabeth Rosenthal. Land use changes now allow grass on a golf course to be labelled a crop and trees planted at a vacation home to be labelled a farm, thus making developers eligible to receive water marked for irrigation.[16]

In Australia, water trading has gone a step further, with both domestic and foreign investors buying up local water rights. In 1994, concerned about the growing strain on the Murray–Darling Basin, the government converted water licences to water rights (called "entitlements" in Australia) in the hope that water trading would lead to conservation and more efficient use of existing water allocations. As in the United States, water trading at first took place among farmers. But, starting in 2000, various governments passed new laws permitting anyone to hold a water right, and allowing private investors, not just smaller agricultural users or landowners, to purchase water rights. This changed the game. Large agribusiness started buying up the water rights of small farmers while governments encouraged bidding wars. In just a few years Australia created an annual $2.6 billion, largely unregulated water market that involves hundreds of brokers buying and selling water on the open market.

The situation is ripe for fraud and embezzlement, say award-winning Australian journalists Deborah Snow and Debra Jopson. "Anyone can hang their shingle up and away they go," says the National Farmers' Federation, which reported on one agent who made half a million dollars in commission in just twelve months. One water broker told the *Sydney Morning Herald* that at one point he had close to $5 million of farmers' money sitting in an ordinary account and could have taken off with it. Some brokers are working for both the buyers and the sellers without telling either party. Some are granting themselves secret commissions, and some are putting buyers' funds into their own private accounts during settlement periods. Snow and Jopson quote Jeff Shand, former chairman of the Australian Water Brokers Association, as saying, "The water trading world is just a cowboy job. Just open slather, in terms of trust accounts."[17]

Big domestic investors are cashing in on this lucrative market. Macquarie Bank, through its subsidiary Macquarie Agribusiness, is scooping up Murray–Darling water rights to secure water for its investment in high-value export-oriented almond crops in Victoria. International investors are also circling the lucrative Australian water market, looking to snap up hundreds of millions of dollars' worth of this resource, with almost no government limit on how much they can buy, say Snow and Jepson in a report for the *Sydney Morning Herald*. The San Diego–based Summit Water Holdings hedge fund (which also owns more than $200 million in water rights in the western United States) bought $20 million worth of permanent rural water rights. Meanwhile, Australian water companies such as the Causeway Water Fund are travelling around the world seeking foreign investors for a "diversified portfolio of permanent water entitlements."[18]

Not surprisingly, since the introduction of water trading, water prices have skyrocketed, going from $2 a megalitre in 2000 to more than $1,500 a megalitre in 2013. This has made it very expensive for the federal government to buy back enough water from private interests to save the Murray–Darling Basin, which has been especially hurt by the recent decade-long drought. The government has pledged billions to buy back the very water it gave away for free just two decades ago — a huge profit for a handful of investors. Cubbie Station, a giant cotton operation located in southwest Queensland, is a case in point. The largest irrigation property in the southern hemisphere, Cubbie Station covers an area the size of the city of Canberra and has acquired enough water rights to more than fill Sydney Harbour, dwarfing all the other irrigators in the country.[19] For years Cubbie's access to so much water has caused dissent among other farmers and environmentalists who believe that the company is seriously harming water flows in the Murray–Darling to grow a thirsty crop that largely gets exported.

In August 2012, the Australian government approved the sale of Cubbie Station to a Chinese-led consortium for only $232 million, a move that created a firestorm of protest over the loss of public control of Murray–Darling water. In March 2013 the new owners started selling their water rights back to the government, earning $47 million from the first sale. With 500,000 megalitres of stored water, the consortium stands to make a healthy profit if it continues to sell its water entitlements back to the Australian people.[20]

Economists around the world tout the Australian market experiment as a model for others to follow. Soon, say some analysts and investment brokers, Australia's water trading market will be global. Ziad Abdelnour, CEO of the New York–based private equity firm

Blackhawk Partners, says that in recent years international law has laid the foundation for the future global trade in water. On one hand, he points out, the World Bank is privatizing water in the Global South, and on the other, international and bilateral trade agreements have created a legal framework for allowing and protecting the sale and trade in water.[21]

It is hard to see, however, who has benefited from the Australian model other than those who have made profits from it. Australia is the driest inhabited continent on earth, but its people have lost control of their scarce water supplies. Even though the Australian constitution defines public rights to water, based on the principle that rivers are common property, the reality is that two decades of market trading have largely invalidated its original intention. Professor George Williams, a constitutional law specialist with the University of New South Wales, says the public trust notion in the constitution is "old and creaky...from the time of paddle steamers." Governments can now legally manage the state's water "in a way that is more about returning a profit than returning a common good," he says.[22]

This is a travesty and a betrayal, says writer and environmentalist Acacia Rose of the Snowy River Alliance, which has fought government-sanctioned over-extraction of the Snowy for years. Water once belonged to the Australian commons, she says; it was owned by the people at large and by the environment. It was available under riparian licence to farmers and to the people under public management of water utilities. "The brave new world of water barons, the globalization of natural resources, the privatization and trading of water, air and carbon means that no farmer can rely on any guarantee for any crop unless he or she pays a premium for high security water."[23] But as lawyer and activist Kellie Tranter concedes, a farmer's loss is an investor's gain. "In the driest inhabited continent on the planet," she says, "the only way for the price to go is up. Those who can afford to pay for the water will have the luxury of it."[24]

PUBLIC TRUST, NOT WATER TRADING

Water trading is a dangerous new trend in the commodification of water and it must be stopped and reversed. Water trading fails to give water priority to municipalities, local farmers, human needs, and ecosystem preservation. Water trading allows governments to abdicate their role of allocating dwindling supplies of water according to the community values they should be upholding, allowing allocation decisions instead to be made based on the ability to pay. In the Canadian province of Alberta, where water trading is in its infancy, the Municipal District of Rocky View paid $15 million to secure water for a $1.6 billion entertainment and racetrack complex development and transfer water from the stressed Bow River. Local farmers and environmentalists were outraged that profit for a development had determined the priorities for the region's scarce water supplies.

Water trading often entrenches the sins of the past. In some jurisdictions water has been extensively over-allocated, resulting in a situation where there are more rights than actual water. Water trading promotes speculation and diminishes the right of the public to know where local water supplies are going. Water trading gives a small group of people and corporations undue control over water sources; it can be dangerous when paired with investor-state agreements that give foreign corporations proprietary right over natural resources. Water trading allows corporate farms and hedge fund investors to take prime farmland out of production, lay off farm workers, and make huge amounts of profit on water that should be safeguarded as a public trust. Water trading can result in permanent dislocations of water, draining aquifers and altering rivers.

The method used many decades ago of allocating water to the first to arrive, as in western Canada and the western United States, does not serve the current reality of a modern world and overtaxed watersheds. Similarly, the move to privatize and allow the sale of water access, as Australia and Chile have done, violates the right to water and removes it from democratic control. Governments that have gone down the water-trading path can and must introduce new laws that take back public control over their water heritage. As our consciousness grows about the human right to water in a water-stressed world, we need to reclaim our water heritage and protect our water commons for all time.

Instead of promoting water trading, governments should be adopting a system of public trust, an important tool in the search for solutions to both the ecological and human water crises. Under a public trust regime, all competing uses for a watershed or aquifer would have to pass a test not just of fairness of access today but also the future capacity of the water body. Public trust offers principles that combine public good, control, and oversight with long-term protection of the watershed. It also sets the stage for an agreed-upon "hierarchy of use" whereby some uses of the water — such as for essential human needs and for ecosystem protection — will take preference over others.

There is a rich history of public trust in U.S. law. The Supreme Court of Idaho has stated that "the public trust doctrine at all times forms the outer boundaries of permissible government action with respect to public trust resources."[25] In 1983 the California Supreme Court used public trust to curtail the diversion of water to Los Angeles from fragile Mono Lake. The Audubon Society successfully argued that even though the tributaries feeding Mono Lake were not navigable (up till then only navigable waters were subject to public trust protection), the public trust was still violated because diverting from those streams jeopardized the public value of the lake.[26]

Two decades later, environmental lawyer Jim Olson used the public trust doctrine to argue for limits to tributary groundwater access, with dramatic effect, in a 2004 court challenge against a Nestlé bottling operation in Michigan. He insisted that groundwater and surface water are one and the same, and that therefore the effects are the same whether the pipe is in a stream or in the groundwater that feeds it. Both must be equally protected for the common good. Olson argues strongly for public trust to be asserted even in the western states, which, he says, in spite of their history of prior appropriation, still own their water.[27]

Olson and others point to the example of some of the eastern states and their assertion of the public good in water governance. In 2008, concerned about major groundwater extractions, Vermont passed a groundwater protection act that declared its groundwater to be a public trust resource, legally belonging to all Vermonters, that must be managed in the best interest of all. A permit system has been set up for those who use over a certain limit per day, and the state has the right to revoke these permits if they are abused. Recently the Vermont Natural Resources Council used the state's public trust legislation to challenge a tritium leak from the Vermont Yankee nuclear power plant, declaring that violation of the integrity of the water was a violation of the rights of the owners — the people of Vermont. Maine has introduced a law that would require a majority vote of the local community before any large groundwater withdrawal or large-scale transport of public water could take place.

Jim Olson recognizes the extraordinary efforts of environmental and conservation groups around the world, as well as regulatory frameworks that have been put in place in many countries to deal with the growing water and other environmental crises. But he says it is time we acknowledged that they are not enough. He feels that something more fundamental — something game-changing, "a shift in paradigm, framework and principle" — is in order. He says the doctrine of public trust could provide a "unifying role" in addressing these issues by establishing outer limits on all government and private-sector actions, and calls for the universal adoption of public trust to protect all aspects of the hydrologic cycle. Olson reminds us that water passes through a complex series of cycles that affect waterways, soil, air, forests, plants, animals, and humans. He quotes the nineteenth-century American jurist Thomas Cooley, who said that because water is a "moveable, wandering thing" it must of necessity "continue common by law of nature."[28]

"For these reasons," writes Olson in the *Vermont Journal of Environmental Law*, "a possible answer is the immediate adoption of a new narrative, with principles grounded in science, values, and policy, that views the systemic threats we face as part of the single connected hydrological whole, a commons governed by public trust principles." He goes on to say:

> The public trust is necessary to solve these threats that directly impact traditional public trust resources... The most obvious whole is not a construct of mind, but the one in which we live — the hydrosphere, basin, and watershed through which water flows, evaporates, transpires, is used, transferred, and is discharged in a continuous cycle. Every arc of the water cycle flows through and affects and is affected by everything else, reminiscent of what Jacques Cousteau once said, "We forget that the water cycle and the life cycle are one."[29]

6

TARGETING PUBLIC WATER SERVICES

My target is to grow the water and waste business to rival the size of our power business within three years. If we get the enabling environment right...we will get more investable opportunities to put real money behind at the end of the day. — **Usha Rao-Monari, World Bank**[1]

A NOTHER THREAT TO WATER as a common heritage is the privatization of water services. Until recently, local water supplies were assumed to belong to local communities. Even as villages became towns and towns became cities, municipal authorities created water systems based on the belief that the local water bounty belonged to all. Public sanitation systems were also considered a key component of public health and a way to stop the spread of communicable diseases. Most water services in the world are still delivered as a not-for-profit public service.

COMMODIFICATION OF WATER SERVICES

Privatization of drinking water and wastewater services was delib-
erately imposed on the Global South by international institutions
and water companies (and their governments) in an open attempt to
capitalize on the growing water crisis in poor countries. It was part of
a development model, beginning in the 1980s, establishing funding
conditions that included privatized essential services in the Global
South.

The choice was clear: if you wanted World Bank funding to provide water services for your people, you had to be open to dealing
with private utilities, most of which were based in Europe. The biggest water utility in the world is Veolia Environnement, with almost
313,000 employees and 2012 revenues of more than $46 billion. The
second biggest is Suez Environnement, which employs more than
80,000 people and had 2012 revenues of $20 billion. Their parent
companies had run the water systems of France for more than a century and were well positioned to move into the Global South when
the World Bank turned on the funding tap for water services.

The commercialization of water services has gone through various phases. Many earlier projects involved a private utility fully
owning and operating the water services, but in many countries
around the world there was a huge backlash against the complete loss
of control over drinking water. Most now promote "public–private
partnerships," or P3s, long-term contracts between public authorities and businesses to design, build, finance, and operate public
water utilities. Proponents claim that P3s leave the public agency in
control of the policies but turn over delivery to a private operator.
Despite the change in terminology, public–private partnerships still
put private companies in control of water services and water rates,
and those companies still need to make a profit. That profit comes
from public funds and from consumers. In an in-depth analysis of

many large-scale public–private partnerships across Europe, the
international NGO Bankwatch Network, which monitors financial
institutions, found that P3s have a long-term detrimental effect on
public budgets and public services alike.[2]

More recently, commercialization of water services has come in
the form of corporatization, whereby a government transforms its
public water utility into an arm's-length publicly listed corporation
so that it can be run on private-sector principles. Corporatization
of water services has been promoted by the World Bank and others
who favour privatization, in countries and municipalities where
either there has been strong resistance to private water services
or such a large population lives in poverty that the private sector
has not been prepared to invest its money. In essence, the public
water utility becomes a state-owned corporation; the government
retains majority ownership of the stock and runs it on the same
commercial basis as a private company, including full cost recovery and profit maximization.

In May 2011 the government of Ireland signed a memorandum of
understanding (MOU) with the International Monetary Fund and the
European Union to reform its water sector to comply with strict new
austerity measures. The government then established a public water
utility called Irish Water, with a clear mandate to operate as a private
company. In exchange for an 85-billion-euro bailout for the country, the MOU required that "the public provision of water services is
to end and this function is to be transferred to a utility company,"
and it further committed Ireland to move towards full cost recovery
through water metering. Irish Water is now a separate subsidiary of
Bord Gáis, the state-owned gas company, and operates as a business.
The company is set to meter homes across the country, and the price
of water is expected to reach as high as 400 euros (approximately
$520) a year for average families. Water resources are abundant
in Ireland, and until the creation of the new utility, water services

for residents had been delivered free of charge. Costs were paid for through tax revenues and by charging commercial users. Irish Water is a clear example of the corporatization of a public water service.

In a 2012 in-depth study of water corporatization, Jørgen Eiken Magdahl, of FIVAS, the Norwegian Association for International Water Studies, explains that this model of water delivery really parallels a private model, and it needs to be put under the same scrutiny by those who favour real public services. The new entity must behave just like a private company, he says, or it will not have the support of the World Bank. He adds that corporatization can lead to privatization, as it is easier to sell a public company that looks and operates as if it were private.

Magdahl studied the impact of this model of water services on several countries in sub-Saharan Africa, where over 50 percent of the population lives on less than $1.25 per day. He found that government practices such as metering and cutting off water supplies were no different than those of private companies, and that the World Bank is still able to promote its "neoliberal ideology" through the use of state power, even in the absence of private operators.[3] South African water activist and researcher Mary Galvin told me that the neoliberal view of water services is so pervasive in South Africa that water issues are seen as purely technical, and most decision makers are not even aware that there are other ways of approaching challenges. This, she argues, is behind the many ways in which the implementation of the right to water goes unfulfilled, even in a system that is technically public.

The results of the commodification of water services have been devastating for those in need of water services they cannot afford. Private companies and their government/corporate counterparts must find a significant margin of profit for the same service that a true public utility provides, so they reduce the workforce, cut services, cut back on pollution control systems, or raise water rates — and usually all of these together. Poor countries have seen their water resources turned into a for-profit commodity for the benefit of foreign investors. And millions have been denied water services because they cannot pay the private rates.

Similar patterns have been seen in North American and European municipalities that have privatized their water services. While they may not have privatization imposed on them by the World Bank, northern municipalities are increasingly cash-strapped and choose to defer costs for needed infrastructure. Investment by a private company so they can accomplish these improvements can seem very sweet in the present. But years down the road, long after the original investment has been returned, the continued high consumer rates guarantee a healthy profit for the company. Food and Water Watch found that in the United States, private water utilities charge 33 percent more for water and 63 percent more for sewer services than local government utilities. Further, the water rates of private companies rise significantly over the years.[4] Yet, in the name of austerity, many European countries are moving to sell off their water utilities, ensuring a legacy of exorbitant water rates for future generations and ongoing profits for water corporations.

MANUFACTURING CONSENT

The two most important global institutions to enlist in this crusade were the United Nations and the World Bank. As early as 1992, at the Dublin Conference, the UN declared water to be an economic good and encouraged user fees, even for the poor. Since then the UN has steadily moved towards a private model of water development, guided by a who's who of water and food corporations and the World Bank. Even the Millennium Development Goals as they pertain to water have been affected by this ideology.

Julie Larsen, in her 2011 report on the role of the water industry at the UN, shows how the private sector has gained influence in
almost every agency that is working on water at the UN. She notes
that there are no government bodies or non-governmental representatives on the steering committee of the CEO Water Mandate, the
platform group set up to assist the private sector in water management. This essentially puts all but big business on the fringe when it
comes to creating UN water management policy. Larsen points to the
group's *Guide to Responsible Business Engagement with Water Policy*,
co-authored with the Pacific Institute, which significantly broadens
the role of corporations into the sphere of public policy and management.[5] "To a large extent," she says, "efforts such as the Guide shift
the discourse away from ensuring that access to water is upheld as
a fundamental human right by governments and the international
community, to legitimizing corporate involvement in the development of global water policy."[6]

At the root of the matter is the fact that corporate involvement
in public policy presents a clear conflict of interest. Corporations
whose business models depend on controlling access to water or gaining entry to new water-service markets cannot uphold the public
interest if it conflicts with their raison d'être and shareholder obligations. Larsen adds that UN publications such as the *World Water
Development Report* are now deeply influenced by private-sector perspectives. An industry-heavy group helped guide and draft sections of
the report on "business, trade, finance and involvement of the private
sector."

The World Bank openly promotes water as a for-profit business
rather than a public trust. Usha Rao-Monari, head of the water sector
for the World Bank's private lending arm, the International Finance
Corporation (IFC), was clear in a recent interview that the World
Bank's role is to help water corporations make money. "The private
sector is looking at water much more now than it has ever done in the

past, and there's a huge pipeline out there," she said.[7]

The World Bank set out to promote a major shift in water policy in the Global South by actively seeking buy-in from non-governmental organizations, think tanks, state agencies, the media, and the private sector in order to manufacture consent for the commodification of water. The World Bank promotes private water services in the South through several of its component agencies: the International Bank for Reconstruction and the International Development Association, which lend money to poor countries based on the condition that the countries adopt a private water-delivery model, and the IFC and the Multilateral Investment Guarantee Agency (MIGA), which encourage private investors to invest in the water sector in poor countries. In the case of the latter, the MIGA actually insures those investors against risks of all kinds, including local political resistance.

The World Bank also administers the International Centre for Settlement of Investment Disputes (ICSID), a powerful international arbitration court initially intended to settle disputes between the private sector and governments over broken contracts. But increasingly the ICSID is being used to challenge the rights of governments to introduce new environmental or health regulations. Cigarette maker Philip Morris International has brought a case against the Australian government, challenging tobacco policies that were put in place to safeguard public health. And a Swedish company, Vattenfall, which controls two nuclear power plants in Germany, is seeking damages related to Germany's decision to phase out nuclear power.

Water companies are using this court to fight governments that try to regain public control of their water services. In 1999, Azurix, a subsidiary of Enron Corporation, agreed to purchase the exclusive right to provide water and sanitation services to parts of Buenos Aires for thirty years. When the Argentine government issued a warning to citizens to boil their water after an algae outbreak, some customers refused to pay their water bills; the company withdrew from the contract and sued the government. A 2007 ICSID tribunal found in favour of the company and ordered the government of Argentina to pay $165 million in compensation. In 2010 the ICSID again ruled in favour of a water company, in a dispute involving the French transnational Suez. This time it was the Argentine government that rescinded the contract, because of concerns over water quality, lack of wastewater treatment, and mounting tariffs. The company has asked for damages of $1.2 billion. When the carrot of persuasion fails, the World Bank uses the stick of financial compliance.

The results of this policy of commodification have been a disaster, says Boston-based Corporate Accountability International in its April 2012 report, *Shutting the Spigot on Private Water*. Even though roughly one–third of all private water contracts signed between 2000 and 2010 have been a complete failure or are in distress, says the watchdog group, the World Bank is giving more money than ever to private water projects. The World Bank's private sector arm, the IFC, has spent $1.4 billion on private water corporations since 1993; in 2013 that amount will jump to $1 billion a year.[8]

Moreover, says the report's Shayda Naficy, the money is going directly to the corporations themselves, bypassing both governments and the World Bank's own requirements for transparency. The World Bank itself is now investing directly, and in the very water companies it is imposing on poor countries. In September 2010 the IFC quietly finalized a 100-million-euro investment in Veolia Voda, the Eastern European subsidiary of the world's largest private water corporation. A similar World Bank investment in the Philippines had a disastrous result, says Corporate Accountability International, which adds that by taking such a large ownership stake in this company, the bank has created for itself a troubling financial incentive to ignore evidence that a private project is failing.

The World Bank also partners with the World Water Council, which runs the World Water Forum; the Global Water Partnership, which promotes private water services in the Global South and is behind controversial policies to promote public financing of private water services; and major business lobby groups such as the World Business Council for Sustainable Development, which was influential in watering down environmental commitments at the original Earth Summit in 1992. Another World Bank partner, Aquafed — the International Federation of Private Water Operators — exists solely to promote the interests of private water operators. With more than three hundred corporate members from forty countries, it includes all the major private utilities and most national associations of private water operators. Although its members fought the right to water for years and are busy privatizing the global water commons, Aquafed was a key player at the March 2012 Sixth World Water Forum in Marseille, and it proudly claims to have contributed to every thematic session at the forum on the right to safe drinking water.

The latest pro-corporate water group to form is the 2030 Water Resources Group (WRG), a public–private partnership launched in 2008. It comprises the World Bank, the World Economic Forum — the organization that brings a who's who of business and government every year to Davos, Switzerland, and has been one of the major players behind economic globalization — and a number of major water corporations, including Nestlé, Coca-Cola, and Veolia. The WRG promises to expand the role of the private sector in water and sanitation services and to influence the political climate on water governance "to enable more market-based mechanisms" in countries around the world, as it said in a 2008 report.

Nestlé chairman Peter Brabeck has been appointed to chair the Water Resources Group, which has already received $1.5 million in funding. This has caused great concern among groups and communities that are fighting to keep local water sources in public hands, as Brabeck can use this position of influence to further the corporate goals of Nestlé and other private water companies.

NESTLÉ SETS GLOBAL WATER POLICY

Bottled water is a highly contentious form of privatization of the water commons. Bottled-water companies establish plants on specific streams, rivers, and aquifers and then mine them mercilessly. They create mountains of plastic garbage, emit an enormous quantity of greenhouse gases in their production, and use huge amounts of energy shipping these bottles around the world. Their ready availability undermines the need to build public water services in poor countries. One bottled-water executive has shamelessly said that, just as cellphones replaced the need to provide land lines in poor countries, bottled water will remove the need to build public water systems.

In spite of the fight against bottled water in North America, the profits of the industry are growing exponentially. Almost 200 billion litres of bottled water were sold in 2011, and the global bottled-water market now stands at close to US$100 billion. By 2015 the industry will be generating more than $126 billion in revenues and will have grown by 20 percent in just five years.[9] This increase is largely due to expansion in Asia, Africa, and Latin America, where water quality is poor but emerging middle and upper classes are able to buy bottled water. Azaz Motiwala, head of an Indian marketing consultant company, says, "I am highly optimistic about the future of the Indian bottled water market. Increasing scarcity of safe drinking water, changing life styles and aggressive expansion by market players may lead this industry to be the next oil industry in the coming decade."[10]

Nestlé, the food and water giant, has annual sales of $91 billion. It is the world's largest bottled-water company and is aggressively marketing bottled water to both the rich and the poor in countries with a growing water crisis. Its wildly successful Pure Life brand is cheap to bottle because it is just purified municipal water. In April 2012 Nestlé announced that it had purchased Pfizer's infant-nutrition unit, which will allow it to sell infant formula alongside Pure Life bottled water to what it calls "less affluent consumers in emerging markets." Wenonah Hauter, of Food and Water Watch, says it is no coincidence that the company will market the two together. Poor women will have to use bottled water rather than tainted local sources to mix their baby formula, an added bonus to Nestlé. Selling water to countries that do not have clean water capitalizes on the water crisis and does nothing to solve it, adds Hauter.[11]

In the United States, Nestlé targets immigrant populations that come from countries where the water is not safe and so do not trust tap water. The company has spent millions on television commercials targeting immigrant Hispanic mothers, claiming that Pure Life promotes nutrition. Most of the target audience are from low-income communities that could save a great deal of money by drinking tap water, which is very safe in the United States.

At the other end of the spectrum, in China the market is upscale. The state of China's waterways, over 70 percent of which are badly polluted, has opened up huge new opportunities for Nestlé, which sells water to wealthy Chinese clients in boutique stores. A Nestlé television ad shows children making unhappy faces after tasting tap water. One child pours his glass into the fishbowl instead of drinking it. His face lights up when his mother offers him a bottle of Pure Life instead. Sales of bottled water in China will climb to $16 billion by 2017, up from just $1 billion in 2000.[12]

Peter Brabeck's positioning of himself as an expert on the global water crisis is of particular concern, given the influence he wields on water policy through his advisory role with governments and the World Bank. It is one thing to capitalize on the global water crisis by selling bottled water, but another to promote privatization of water services and water trading from a position of power. The man who started as an ice-cream salesman for Nestlé now also influences the water aid policies of the Swiss government, through the controversial Swiss Water Partnership of that country's Federal Agency for

Development and Cooperation. Nestlé is also a charter member of the UN's CEO Water Mandate, and has been influential within the powerful UN agencies that shape water policy.

So it matters a great deal to the future of water in the world that Peter Brabeck avidly promotes private control of water. He once famously described the notion that water is a human right as "extreme," a statement that was widely criticized. He now admits that there is a need to set aside some water for the most needy, but he says that the market should determine the fate of the rest of the world's water: "Give the 1.5 percent of the water, make it a human right. But give me a market for the 98.5 percent so the market forces are able to react, and they will be the best guidance that you can have. Because if the market forces are there, the investments are going to be made."[13]

Brabeck supports water markets and water trading. He created a storm of criticism in Alberta, Canada, in July 2011 when he announced that he was in talks with the Alberta government to establish a "water exchange" to allow water to be traded and sold as a commodity. He added that the province is ideal for such a scheme because water in Alberta is scarce and the competition for it fierce. He sits on a board that advises the University of Alberta on water policy, along with top names in the energy industry in a province where tar sands energy operations are destroying local water sources. Students, farmers, environmentalists, and others who are trying to retain democratic control of Alberta's declining water supplies picketed Brabeck when he received an honorary doctorate from the university.

Brabeck has embraced a concept called "creating shared value," first put forward by the *Harvard Business Review* to promote corporate social responsibility. At a time of public distrust of big business, the idea was to merge what is good for business with what is good for the community in a win-win scenario. Brabeck promotes the notion that what is good for Nestlé, which he calls "the leading global health and wellness company," is good for the world, and that together, all "stakeholders" — business, government, and civil society — can create consensus on water policy.

One key Nestlé "shared value" is the need for conservation of the world's water, a theme Peter Brabeck often speaks on. This has clearly not been transmitted to Nestlé Waters Canada. In October 2012 the company appealed a decision by the Ontario Ministry of the Environment to impose mandatory reductions on water takings in times of severe drought; Nestlé has a lucrative water-taking operation in Guelph, Ontario. Another stated Nestlé value is compliance with the law and codes of conduct. However, in January 2013 a civil court in Switzerland found the company guilty of spying on ATTAC Switzerland, a social justice organization critical of Nestlé that was gathering information on it for a publication. Nestlé was ordered to pay damages to each of the nine victims.

In 2009, a group that includes Corporate Accountability International, the International Labour Rights Forum, trade unionists from the Philippines, and Baby Milk Action called for Nestlé to be expelled from the UN Global Compact for trade-union busting, the use of child labour in Colombia, and environmental degradation of water resources in many parts of the world — all activities that violate the company's corporate code of conduct.

Peter Brabeck uses his role with the World Bank's Water Resources Group to promote commodifying of the world's water. The group's strategy is to insert the private sector into water management, one country at a time, through a combination of industry-funded research and direct partnerships with government agencies, reports Corporate Accountability International. In order to be eligible for funding, all projects must provide for at least one partner from the private sector. This violates the World Bank's own goal of poverty alleviation, its stated commitment to the rights to water and sanitation, and its rules on transparency.[14]

At the 2010 World Economic Forum, the Water Resources Group launched a series of pilot projects with a clear goal: "to build a demand-driven public–private platform to support governments that want to engage in water sector reforms."[15] Given that most developing countries will have no choice if they want World Bank funding, it is disingenuous to imply that the Water Resources Group is doing those countries a favour by setting up these projects. Any country that needs funding for water services is letting not just the World Bank into the inner circle of government but also Coca-Cola, PepsiCo, Suez, Veolia, and, of course, Nestlé. Sadly, this initiative is aided by public funds through the development agencies of Switzerland and Germany.

The model is clear: governments and their citizens put up the money, the private water sector delivers the water services for profit, and friendly NGOs provide charity to the poor — the exact opposite of a model of water justice and democracy. The "shared value" of water privatization is imposed on poor countries and adopted with enthusiasm by rich countries and some developing countries, such as South Africa and India, whose governments have adopted the market model of development. The concept is also being used as a weapon against local activists who do not buy into a corporate future for their local water systems. And communities around the world are feeling the effect of the loss of their water systems.

7

THE LOSS OF THE WATER COMMONS DEVASTATES COMMUNITIES

Water is life and depriving anybody of this natural resource is nothing less than depriving somebody of the right to life. Snuffing out the life of a person by intent is tantamount to murder.

— Moulana Usman Baig, All India Imams Council

INDIA'S KARNATAKA: PETRI DISH FOR WATER PRIVATIZATION

One of the first Water Resources Group projects is in the water-starved state of Karnataka, in the southwest of India. So serious is the water crisis there that 7,500 villages are experiencing a severe water shortage and another 15,000 are at risk. Scientists are warning that the state could actually run out of water. Further, 80 percent of the population do not have potable water in their homes and 68 percent practise open defecation. To alleviate the crisis, a number of things are desperately needed. Rainwater harvesting and watershed restoration would renew the hydrologic cycle, while a program of democratic local management of watersheds and a plan to provide water and sanitation for the population, as required under the

UN resolution, would allocate resources to the benefit of the whole population.

However, Karnataka and India have adopted aggressive policies to privatize India's water resources in order to entice foreign capital to help transform India from a still largely rural society to an urban, industrialized one. The Indian government has targeted a model of sustained annual economic growth of 8 to 10 percent. This has created a pattern of development in India, says Madhuresh Kumar, of the National Alliance of People's Movements, that sees India's water as a resource for industrial development. India is facing a reduction in its clean water supplies caused by intense exploitation of groundwater, industrial and mining activities, the construction of big dams and thermal power plants, and increased water contamination from extractive industries. Millions of people have been displaced.

With the water table falling at the same time that demand is escalating, "a credible scenario would be the diversion of freshwater sources to industrial purposes ... putting human welfare at great risk," says Kumar.[1] In other words, the time may come when the needs of the population may be pitted against the needs of development, and governments will have to make a choice. The more water in private hands, the more likely the choice will be to promote water for economic growth. If this seems harsh, one needs only to look at the millions already displaced to make way for big dams and free-trade zones to see the emerging trend.

This emphasis on using the country's water to promote economic development is mirrored in the long-awaited update of its water policy. While India's 2012 National Water Policy pays lip service to water as a community resource, it also refers to it as an economic good, to pricing on economic principles, and to public–private partnerships. It disappointingly does not recognize water and sanitation as basic rights, and allows for water infrastructure to be built with public funds but run by public–private partnerships, as advocated by the

World Bank — just the model the Water Resources Group is promoting in Karnataka.

State authorities and business leaders, who call Karnataka the "Silicon Valley of India" and its biggest city, Bangalore, the "call-centre capital of the world," have targeted the state for industrial development and have put its water on the market. The 2002 State Urban Drinking Water and Sanitation Policy, written with advice from the World Bank, marked a departure from the traditional view that water is a commons whose access is a fundamental right to the view that water is a commodity, says the People's Campaign for the Right to Water. The policy ensures universal coverage of water and sanitation services "that people want and are willing to pay for."

The state became a poster child for water privatization, dismantling its public water distribution. Veolia and other private, for-profit utilities now run water services in a number of cities for those who can afford them; shockingly, thousands of public taps have been removed. This has had a devastating impact on the millions of people dependent on those taps. Bangalore's government has announced that it will privatize its water, which has met with fierce resistance. In February 2011 the U.S. Commerce Department sent a water-trade mission of business executives to Bangalore to "tap the $50 billion Indian water market" and assist American companies to "seize export opportunities in this sector," in the words of one of the delegates. The United States has targeted Karnataka because the state is recognized as a leader in dismantling public systems for water distribution, says Kshithij Urs, of ActionAid in Karnataka. The trade mission has changed the Indian situation completely, he adds.[2]

Karnataka is one of the experiments of Peter Brabeck's Water Resources Group. The Indian partners are not local governments or councils but the Confederation of Indian Industry and the Council on Energy, Environment and Water, a public–private lobby group that has partnered with the utility giant Veolia on urban water renewal. The stated purpose of the project is "to help government develop a water action plan for the transition from an agricultural to an industrial economy," ensuring that Karnataka's ever-dwindling water supplies will "run uphill to money," as they say in water-conflicted California. And millions more will suffer.

PROFITEERING FROM NIGERIA'S WATER CRISIS

Another example is Nigeria, where the water crisis is one of the worst in the world. Well over half the population — 70 million people — have no access to clean water, according to the Ministry of Water, and two-thirds have no access to sanitation. Only China and India have larger populations without water access. In Lagos, due soon to become the third largest city in the world, only about one-quarter of the city's 21 million people have access to piped water. Private tankers, carts, boreholes, and wells supply the rest. Families spend half their household budgets on buying water.

Instead of coming up with a plan to help the country deliver safe public water to its people, the World Bank's International Finance Corporation has repeatedly promoted privatization as the sole solution. In 1999 it developed a proposal that "required" the Lagos state government to seek private-sector operators as a condition of receiving World Bank funds. In 2003 it suggested developing privatization of water in Nigeria through a "franchising" structure similar to the method used for fast-food chains such as KFC, whereby private water vendors would be "branded" by a transnational water company. In 2005 it proposed a $200 million "water sector reform" project based on public–private partnerships. However, there was a catch. Because of the poverty of the country and therefore the limited opportunities to make a profit, no major private-sector corporations would touch any of these projects; no private investment has taken place in the country's water sector.

Nevertheless, reports the Public Services International Research Unit, in 2004, anticipating the promised foreign capital, the Nigerian government passed a law to promote privatization of its water services and gave the Lagos Water Corporation a mandate to start operating on a full cost-recovery basis. The company then started cutting off those who could not pay their water bills and even disconnected the water supply from public primary schools that had defaulted on payment of their bills.[3]

In January 2013 the International Finance Corporation, now advised by Nestlé's Peter Brabeck, restarted talks with the Nigerian government and the private sector to come up with a plan. After years of hearing that the only solution to their water crisis is privatization, the Nigerian government has no alternative plan. Sarah Reng Ochekpe, Minister of Water Resources, said, "Despite the fact that water is a social commodity, we need to look at the economic potential of the commodity."[4] One can only imagine the terms the government will accept to finally attract the private funds so long promised by the World Bank.

Meanwhile, Nestlé is making huge profits from Nigeria's water crisis by selling its Pure Life brand. The company, infamous for peddling infant formula to the poor, now peddles its "creating shared value" project to Nigerians. It holds university and community seminars on "sustainable investment in water and nutrition" and "training the trainers" sessions for public school teachers on "healthy hydration." Says Etienne Benet, head of Nestlé for Central and West Africa, "As the world's leading Nutrition, Health and Wellness company, we go one step further by encouraging people to adopt healthier diets, whether the problem is deficiency in vitamins and minerals at one end of the spectrum, or obesity at the other. This is what we call Creating Shared Value and this is how we demonstrate our commitment."[5]

Nestlé Waters Nigeria doesn't stop with seminars about the nutritional value of bottled water. It directly equates its growing bottled-water sales with helping meet the UN Millennium Development Goals on water and sanitation. Nestlé boasts that its water plant in Agbara is one of the company's intervention strategies for provision of safe, affordable drinking water in Nigeria and "represents Nestlé's contribution to the attainment of one of the Millennium Goals in Nigeria, of which the provision of safe drinking water is a key objective." The company is proud and sees no contradiction in the fact that it has "leveraged its healthy positioning to become the preferred bottled water brand for the on-the-go, out-of-home, and family-in-house consumption occasions."[6] Not surprisingly, given the dismal state of fresh water in Nigeria, the bottled-water market is very lucrative and growing, according to the company, with "excellent long term prospects." In October 2012 Nestlé Nigeria (including both its food and water divisions) posted a 41 percent pre-tax profit increase from the year before.

AUSTERITY PLAN PUTS EUROPE'S PEOPLE AND WATER AT RISK

Europe has a strong history of public water services, although in recent decades many countries have partially privatized some water utilities. But even this limited experiment has brought major problems, and many municipalities are not renewing private contracts. Even France, where private companies have run water utilities for more than a hundred years, has brought water services under public management in many cities, including Paris.

But the financial crisis of 2008 and the austerity agenda of the European Commission (EC) have put public water in Europe on the chopping block. Private banks, which were themselves largely responsible for the financial crash, are strongly behind the push to privatize Europe's water, as are European corporations that stand to profit from the transfer of water services from public to private control. The irony is, of course, that the banks were bailed out with public funds and the public subsidizes the water companies as well.

Together with the International Monetary Fund and the European Central Bank, the EC has imposed bailout conditions for indebted countries that include the sale of public services, including water. In a September 2012 letter to several groups concerned about the imposition of water privatization as a condition of debt relief, an EC official openly confirmed this policy: "The Commission believes that the privatisation of public utilities, including water supply firms, can deliver benefits to the society when carefully made." The official also noted that the privatization of public services would contribute to reduction of public debt.

Water privatization flies in the face of both public opinion and empirical evidence. A 2012 Public Services International Research Unit (PSIRU) report shows not only that many European municipalities are trying to undo failed privatization experiments, but also that many European private water operations end up tapping the public purse for finance and investment. According to the study, private companies have received nearly 500 million euros in financing from the European Bank for Reconstruction and Development since 1991. In fact, both the major private utilities, Suez and Veolia, are increasingly dependent on state capital for their activities in water, says the report.[7]

A Blue Planet Project report on implementing the right to water in Europe, edited by PSIRU's David Hall and the Project's Meera Karunananthan, tells of water privatization failures in countries across the continent. United Utilities of Great Britain has been running the water system of Sofia, Bulgaria, since 2000. It was brought in to stop leak rates of 60 percent, but has not met any of its contractual targets. The city's leakage rate remains at 60 percent, yet water tariffs have risen steadily. The new austerity measures allow for expanded privatization in other Bulgarian towns and cities, as well as steep private-water rate hikes. At the same time, commercial water rates have dropped tenfold in order to stimulate the economy. According to the environmental group Za Zemiata, one thousand households in Sofia were cut off from their water services in 2011 and five thousand more put on notice in 2012 for not paying their fees. Three hundred and seventy families in Sofia were evicted in early 2012 for lack of payment of their water bills. As well, people living in informal settlements are completely neglected by the water company.[8]

Greece's two biggest cities, Thessaloniki and Athens, allowed partial privatization of their water companies, EYATH and EYDAP, in the late 1990s but maintained majority public control. However, under the EC austerity plan, the two cities are set to fully privatize their water services, likely to Suez. Since the beginning of the privatization experiment, the workforce has been cut by almost two-thirds, the price of water services has risen 300 percent, and services have not improved, reports the unions for the two companies. Thessaloniki's EYATH serves a 2,330-kilometre network and 510,000 metres of pipe with just eleven plumbers, reports the union. It predicts higher water rates, more privatization, and more cut-offs for those unable to pay. Already, with a 40 percent loss of disposable income in five years and an official poverty rate of 21 percent, many Greeks have had their power and water services cut. With a government warning that it may run out of money to pay pensions, the loss of a public water system will spell disaster for millions in the "cradle of Western civilization."[9]

Portugal too has a history of public water management with governments promoting privatization, starting in the early 1990s. Rates skyrocketed so that customers pay 30 percent more than those served by public systems. Portugal's austerity program instituted deep cuts to wages, pensions, and social security, and income loss could run as high as 50 percent, says a coalition of social justice and anti-poverty organizations. Despite this, the government has been ordered to sell Agúas de Portugal, the national water company, along with a number of other public service providers, in order to receive debt relief. This leaves the government totally devoid of any effective capacity to develop public policy to deal with the mounting social crisis. At the same time that the minimum wage has been frozen, there has been a huge increase in water rates that has resulted in widespread cut-offs. To add to this travesty, the government ordered all public drinking fountains closed in order to "protect" the profits of the private water companies.[10]

Spain began a process of partial water privatization in the mid-1980s. About 50 percent of Spain's water services are now run by private utilities, most by Agbar, which is owned by Suez, and Aqualia, which has links to Veolia. Madrid has retained a strong public water service with a good reputation but is now planning to privatize Canal de Isabel II, its public water utility, as part of its austerity program. Local citizens are outraged that there has been no transparency or public consultation. Barcelona has handed over its drinking-water services, without open tender, to Agbar. A local citizens' coalition against privatization, Aigua és Vida, says that the government is selling the utility at a fire-sale price in order to meet its debt payment. The group estimates that Agbar will earn 1.4 billion euros over the fifty-year life of its contract, money that will not be put back into source protection or public services.[11]

As in Greece and Portugal, the Spanish people are suffering. Average disposable income has dropped by 10 percent since 2008 and many can no longer pay water, heating, or electricity bills. Anti-poverty groups report hundreds of evictions every day across the country as families fail to make the rent. Homelessness is on the rise. The very worst thing the country can do is remove public services such as water and hand over the country's water supplies to foreign-controlled transnational companies.

While technically the water networks in Italy are publicly owned, governments have passed laws over the past fifteen years to begin the process of privatization. As a consequence, about half the population gets its water services from a public–private partnership. The Forum Italiano dei Movimenti per l'Acqua reports that water rates have climbed along with the process of privatization, up 61.4 percent between 1997 and 2006. In those same years, investment in the water sector declined by more than 70 percent and the water workforce was cut by 30 percent. Lower investment has resulted in deterioration of service as well as environmental damage from poor wastewater practices. This has had serious consequences for local rivers and the Mediterranean Sea. Private boards run most of the utilities with no public transparency.

Public resistance to water privatization has stopped the EC push for further privatization in the water sector for now, but this struggle is far from over. As in other countries, deep cuts to wages, pensions, health care, and social security have had a traumatic effect on the population. Many cities face bankruptcy and cannot maintain their public services, including their water.

MARGARET THATCHER'S LEGACY

Europeans can learn a lesson in water privatization from the place that first embraced it. Great Britain's water rates paid by citizens and the profits earned by corporations have both soared since Margaret Thatcher introduced water privatization in 1989. Twenty-one private companies now run all the water services in England and Wales, and they made a pre-tax profit of more than 2 billion euros in 2011. Household bills, which soared 147 percent in the first decade of privatization, are almost 400 euros a family. The companies now want to install a water meter in every home so that residents will pay the full cost of water, including shareholder dividends. Trade unions and many others strongly oppose this move, citing the damage already done to the poor, the elderly, and the disabled by high water rates.

The companies are also infamous for their record of pollution. Ofwat, the body that oversees regulation of the privatized water system, admits that well over 3 billion litres of water leak out of the country's creaking sewage system every day. But targets to reduce leakages are voluntary, and the agency admits that only eight of the twenty-one private water companies have set a target of zero leaks.[12] Further, Thames Water, which runs London's water system, has been dumping semi-treated and untreated waste into the Thames River for years. For more than a decade the company has been promising to build a state-of-the-art tunnel to take the sewage out to sea (thus merely hiding the pollution in the ocean) but has not done so.

On top of this, the water companies also pay little or no taxes. The *Guardian* reports that Thames Water and Anglian Water paid no tax on their profits in 2012 while generously rewarding their executives and investors. In its March 2013 year-end report, Thames Water again acknowledged that it paid no taxes and had pocketed a government credit of $7.5 million. Both companies have made hundreds of millions of pounds in operating profits, and some have awarded their senior executives with hefty bonuses and huge dividends. Martin Baggs, the chief executive of Thames Water, which enjoyed a tax rebate in 2012 of £76 million (US$116 million) despite making operating profits of £650 million, was given a bonus of £420,000 (US$640,000) on top of his £425,000 salary. In 2013 his salary was raised to the equivalent of US$675,000 and his bonuses came to US$960,000. Baggs is in line for a further windfall of $1.5 million, based on company performance through to 2015.[13]

Observer columnist Will Hutton calls London the "effluent capital of Europe" and reports that, since 2006, Thames Water has been owned by a group of private equity funds domiciled in Luxembourg and marshalled by the Australian bank Macquarie. "By maxing out on debt, all the astonishingly high interest payments can be offset against tax, so that in 2012 it paid no tax whatsoever even while paying £279.5 million of dividends — subject, of course, to minimal Luxembourg taxation."[14]

Small wonder a recent poll found that 72 percent of Britons want their water services back in public hands. Yet a new bill before the British Parliament would actually reduce current restrictions on mergers and acquisitions in the water sector, making it easier for the water companies to merge and easier for businesses to compete in supplying water to England, Scotland, and Wales.[15]

CHILE HANDS OVER ITS WATER ENTIRELY TO CORPORATIONS

Chile has gone further than any other country in commodifying water and creating a market economy based on private water rights, privatization of water services, and damming and corporate control of rivers for mining and other extractive industries, reports the highly respected environmental network Chile Sustentable. In a 2010 report, multiple authors and experts trace the process back to the dictatorship of Augusto Pinochet.[16]

The commodification of Chile's water started with the 1981 Water Code enacted by the military regime, which was based on a strong pro-business bias. For the first time in Chile's history, land and water were separated to allow for the unconstrained purchase and sale of water. While water was defined as a national public good, it was also defined as a market asset, allowing the privatization of water through free and perpetual grants of rights to big corporate interests. Once the water rights are granted, the state no longer has the power to intervene; the reallocation of these water resources is done through the buying and selling of water markets.

The process has been an unmitigated disaster for Chile's people and for ecosystem health. It has led to concentration of ownership of Chile's water among a handful of corporations, many of them foreign and the majority in the export sector. Seventy-five percent of all mineral production is in the hands of private companies, most of them transnational. Three companies own 90 percent of the water rights for power generation nationwide. The Spanish power company Endesa, recently acquired by the Italian state company Enel, controls over 80 percent of the total national water rights for non-consumptive use (water that is returned to the watershed). The agriculture sector consumes close to 85 percent of all water granted for consumptive use (water that is not returned to the watershed), nearly 20 percent of which is exported in the form of virtual exports. All the agribusiness exporters are privately owned.

This concentration of power over water in largely transnational corporate hands has led to an unprecedented assault on the country's surface and groundwater sources. This in turn is causing ecological strain in many areas and creating tensions between local communities and the corporations. The 2010 Chile Sustentable report cites "degradation of the country's most important watersheds" and subsequent shortage of drinking water in many rural villages and indigenous communities. Corporate water consumption grew by 160 percent between 1990 and 2002 and has continued apace. Chilean government figures predict an exponential increase in commercial water use in the coming decade.

Chilean agribusiness uses large amounts of pesticides, herbicides, and fertilizers, all of which degrade watersheds. As well, they drain local water sources to produce export commodities. Communities in the northern town of San Pedro are in a fierce struggle with several agribusiness companies that control water rights in the Yali aquifer. The companies currently consume most of the water the aquifer produces, forcing sixteen local communities to truck in water, and are seeking further rights that would exceed the aquifer's production capacity.

Hydroelectric development by transnationals is threatening protected and indigenous areas all through Chile's southern region, where the water supplies are more plentiful. A Norwegian state company is planning to dam fifteen major rivers; an Italian–Chilean consortium seems poised to build two huge hydroelectric projects inside Puyehue National Park, in the Los Lagos region; and HidroAysén, owned by the transnational Endesa, is seeking the right to build six dams deep in Patagonia, flooding as much as 7,500 hectares of virgin jungle and compromising ten state-protected areas and twenty-six wetlands.

The mining industry is causing "critical deficits" of water in some regions. In the water-scarce Antofagasta region, for example, mining uses more than one thousand litres a second, and mining companies hold almost 100 percent of the groundwater rights. The Chilean Copper Corporation itself reports that this region will experience an "extreme deficit" in drinking water by 2025. They even use case studies to describe the growing conflict over water, since the government is now allowing mining and drilling in national parks and is issuing more water rights than there are actual water reserves, resulting in drying up of areas of the country. Whole water basins have been polluted and/or depleted by mining corporations operating in the absence of regulations.

Because the mining code gives any company the right to dig on any land regardless of ownership, indigenous and local communities have been overrun and exploited by rapacious foreign mining companies seeking ever new mineral deposits. In 1940 there were four hundred families in the northern village of Quillagua, and they had access to 660 litres of water per second from the Loa River. Today only a hundred families live there and they have to share 90 litres per second, all that is left after two mining companies contaminated and diminished the river. Cancer and respiratory problems plague the residents of Chañaral after decades of toxic dumping by the state mining company. The locals in Copiapó, another northern town, remember a time of abundance before two of the three local rivers, the Copiapó and the Salado, disappeared because of over-extraction by mining and agribusiness companies.[17] These stories are repeated all over the country.

In its commitment to a private water model, Chile has privatized its entire water and wastewater sector, contracting with transnational water utilities such as Suez and foreign investment consortia such as the Ontario Teachers' Pension Plan. (Ontario teachers have been criticized for allowing their public pension funds to be used to privatize essential services in other countries but there has been little debate among the teachers themselves about this practice.) Chile's water sector has proven to be a windfall for these companies and their investors, many of whom make steady annual returns of 25 percent. The government subsidizes these profits by guaranteeing the water companies a return of at least 10 percent and paying to provide water service in some poor households. The citizens of Chile subsidize these profits with high water rates. As a result of privatization, Chile's water rates are now the highest in Latin America. Water consumption in households has dropped as a direct result of these high rates, even as the government continues to subsidize exploitation of water by the private sector.

Jobs have been lost since privatization, and rate inequality has been documented in different parts of the country. Importantly — to refute those who say that privatization brought the miracle of water services to Chile — privatization has not meant an improvement in coverage or access to water resources for Chileans. The percentage of the population covered by drinking water and sewage services was almost exactly the same before privatization as it was ten years later, in 2008. The only area where there has been improvement is in the treatment of waste water, which is paid for by the consumer at steadily rising rates.

Thirty years after its experiment with water privatization, Chile is faced with serious issues of violation of basic human rights, growing conflicts over declining supplies, water insecurity for the future, and environmental degradation. This is manifestly a failed experiment.

8

RECLAIMING THE WATER COMMONS

Water is life. The People's Water Board advocates for access, protection, and conservation of water. We believe water is a human right and all people should have access to clean and affordable water. Water is a commons that should be held in the public trust free of privatization. The People's Water Board promotes awareness of the interconnectedness of all people and resources. — **Detroit People's Water Board mission statement**

THE ENCLOSURE OF THE water commons is destroying communities and denying people the right to water and sanitation. This right cannot be fulfilled under a market or corporate model of water governance. To ensure that the world's dwindling water sources are more equitably shared, we must promote the values of conservation, water justice, and democracy. If we continue to allow commodification of the world's water, it will flow towards those with money and power, not to the communities and ecosystems that need water for survival.

Since writing *Blue Gold* and *Blue Covenant*, I have watched the global water crisis deepen as more people and more money chase the declining reserves of water. I have watched governments and international institutions turn over their control and decision-making powers over the world's water supplies to the private sector, negatively affecting their own ability to manage these supplies in the public good. I have become convinced that if we do not declare water to be a common heritage and a public service and protect it in law as a public trust, we will see the day when governments will no longer have the ability to provide water for people or to protect water for the planet.

A MOVEMENT COMES OF AGE

I have also had the honour of being part of a grassroots movement that includes thousands of communities and networks that have come together to fight for water protection and water justice. One of the first water justice networks was Red VIDA, a coalition that came together in 2003 to fight water privatization in Latin America. The famous water war in Cochabamba, Bolivia, was still fresh in people's minds. In the mid-1990s, at the direction of the World Bank, Bechtel set up a private water subsidiary in Cochabamba. When it tripled water rates and charged for the rainwater people collected, the population of this largely indigenous state rose up and defied the army, forcing the company to retreat. All over Latin America, groups were forming to fight Suez and Veolia (then Vivendi), which were newly operating in their communities. The citizens of Uruguay were preparing for a successful referendum to amend their constitution to recognize the right to water, and activists in Colombia, Mexico, and other countries were preparing similar strategies.

Working with environmental groups such as Friends of the Earth International, Food and Water Watch, and Via Campesina, the international peasant movement, thousands of water activists met at consecutive World Water Forums in The Hague, Kyoto, Mexico City, Istanbul, and Marseille and openly opposed the World Water Council. We gathered together at World Social Forums from Brazil to India to Tunisia, where we shared information strategy, resources, and support. We were tear-gassed at the World Summit on Sustainable Development, RIO+10, in Johannesburg in 2002, and marched together at its successor, RIO+20, in Brazil ten years later. Along the way we set up networks in Africa, Australia, Asia, Europe, and North America.

A cherished partner has been Public Services International (PSI), a global federation of public-sector unions representing 20 million working women and men in 152 countries. PSI champions human rights, advocates for social justice, and promotes universal access to quality public services. The indomitable David Boys co-ordinates PSI's global water campaign; he has been a tireless fighter for both public water and public-sector water workers. He has marched with activists at every one of these global events. Whether it is EPSU in Europe, CUPE in Canada, COSATU in South Africa, or the Alliance of Government Workers in the Water Sector in the Philippines, the partnership between water justice advocates and public-sector unions has provided the backbone for this powerful movement.

PSI is also the major supporter of PSIRU, the research institute located at the University of Greenwich (U.K.) that carries out empirical research into globalization and privatization of public services. Until his retirement in August 2013, it was headed by David Hall; without him and his team it would have been next to impossible to build a case for public water based on independent evidence. When I look at what we have accomplished in a mere fifteen years, I realize the power of an idea whose time has come.

RECLAIMING PUBLIC WATER

Communities around the world are struggling to keep their water systems in public hands or to bring private systems back under public control. While privatization has spread throughout the world over the past few decades, the tide does seem to be turning. A website called *Water Remunicipalisation Tracker* is maintained by Corporate Europe Observatory, a research and campaign group that exposes corporate abuse, and the Transnational Institute, which focuses on international justice issues. The website reports on the progress of efforts to reclaim public water services around the world. "A major trend has emerged as more and more communities insist on returning water and wastewater services to public management through remunicipalisation, forcing water multinationals to pull out of services in Latin America, the United States, Africa and Europe," the groups state.

Some of the most intense confrontations have taken place in Latin America, where the privatization experiment of the Global South was incubated. The fight against private water projects in Cochabamba and La Paz in Bolivia, in Buenos Aires and Santa Fe in Argentina, and in Uruguay were successful. Similar battles have taken place in Africa, where both Mali and Dar es Salaam, Tanzania, have taken back their water services from private operators. In Asia, activists turned back water privatizations in Jakarta, Indonesia, and in Malaysia. Cities in Ukraine, Georgia, Kazakhstan, and Uzbekistan have followed suit.

Europe has seen the largest number of privatization reversals, the most striking in France, where water privatization was born. In the past decade more than forty municipalities, including Paris, Toulouse, and Nice, have brought their water service operations under public control, many for the first time. Paris saved 35 million euros ($47 million) in its first year as a public water provider and was able to reduce water prices to consumers by 8 percent. This has been a major blow to Veolia and Suez, striking at the heart of their operations.

Despite the intention of the government of India to promote private water utilities, there has been fierce opposition across the country. In June 2013, after years of its being managed by foreign corporations (including Suez), Jakarta repurchased its water service, which will now be operated as a public agency. It was a major victory for the Coalition of Jakarta Residents Opposing Water Privatization, a network of activists and communities that have worked for sixteen years for this decision.

Over the past decade, say PSIRU's David Hall and Emanuele Lobina in a May 2012 report on multinational water companies, the big water transnationals have in fact decided that most international expansion should be abandoned. In 2003, the report states, Suez announced that it would withdraw a large proportion of its contracts and not take on any new ones without a secure high level of return. Veolia's withdrawal has been more erratic, but in 2011 it announced that it would leave nearly half of the seventy-seven countries in which it was doing business. So vulnerable have many private water companies become that they are more dependent than ever on the support and partnership of development banks and governments. A national strategy to protect key French companies from foreign control resulted in the French state's becoming the major shareholder in Suez, Veolia, and SAUR, the third largest French water company.[1] The profits of all these companies dropped in 2012 and are predicted to continue falling until at least 2014.

In the 2012 book *Remunicipalisation: Putting Water Back into Public Hands*, the Municipal Services Project, an international research project exploring alternatives to privatization and commercialization of public services, studied a number of remunicipalizations around the world. As the book reports, this trend shows that the public system can outperform the private sector and can be an effective water provider anywhere. In his chapter, Canadian academic David McDonald, professor of global development studies at Queen's University in Kingston, Ontario, shows that this debate has been fought before as he charts the history of public water services in London, England. That city's water works started as a patchwork of private utilities in the 1850s, became a public monopoly in the 1900s, and then reverted to a patchwork of private services in the 1990s. McDonald quotes Joseph Chamberlain, the mayor of Birmingham in the 1870s, who declared, "The Water Works should never be a source of profit, as all profit should go in the direction of the price of water."[2]

McDonald writes further that while the team's case studies come from all over the world and are very different, they all demonstrate that water services can be transferred from private to public ownership and management with little disruption of service and positive results. Significant direct savings for consumers and systemic efficiency gains can be attained through good transparent public management. Short-term savings result in long-term infrastructure development, and staff morale improves under a public system.

Despite these successes, Denmark's Jørgen Eiken Magdahl of FIVAS reminds us that we must be vigilant about ensuring that we do not mistake corporatization for remunicipalization. If the "public" water service is run on the same for-profit basis as a private company, this is not a victory. As well, we may need to revisit what we mean by "public water." The process should not be seen as a polarization of private versus central state delivery, but rather as an opportunity to rethink how we define successful water services.[3]

PUBLIC–PUBLIC PARTNERSHIPS

Local control and co-operation are key to the success of an alternative to P3s called public–public partnerships (PUPs), in which two or more water utilities or non-governmental organizations pool resources to buy power and technical expertise. Food and Water Watch Europe says that communities often cannot afford to maintain their own drinking and wastewater systems. By partnering with other public entities, they can avoid the problems that plague strictly private operations. And because they do not involve investors who expect a cut of the savings, efficiencies generated are reinvested in the system.

"For communities in the developing world, these partnerships can serve as the foundation for sustainable economic development," says a 2012 report, *Public–Public Partnerships: An Alternative Model to Leverage the Capacity of Municipal Water Utilities.* "PUPS between water systems in industrialized countries and developing countries...improve water quality in the developing world by sharing best practices." This model has been particularly successful in Africa, where more than half a dozen cross-border utility partnerships have been forged since 1987. A European Commission project has funded a number of regional water co-operatives, such as western Kenya's Rift Valley Water Services Board, which has overseen an expansion of piped water and storage in a region that holds 5.5 million customers.[4]

Queen's University's David McDonald and Gemma Boag, a water scientist at Oxford University, studied thirty public–public partnerships around the world and found that they had several characteristics in common. PUPs improve capacity at minimal cost, promote democratic and equitable community water services, build solidarity, bring back a focus on public services to decision makers, and beat back privatization by confounding the myth that the private sector can do it better.[5]

In a report of "exemplary cases" of public water management in Europe, the Transnational Institute and Corporate Europe Observatory point to public water companies in Vienna, Munich, and Amsterdam that provide for millions of people while proving that ecological responsibility can be achieved at relatively low cost, once the profit motive is removed. They highlight the participatory water management systems of Cordoba in Spain and Grenoble in France. They praise the Turkish municipality of Dikili, which has introduced a socially responsible approach, cancelling debts for unpaid water bills and providing a minimum amount of free water. The authors then outline the criteria for successful progressive public water management: good quality, universal service, effectiveness in meeting needs, social equity, solidarity, sustainability, good working conditions, democratic structures and control, and progressive legislation.[6]

And communities around the world are fighting to promote this model of a public water commons.

EUROPE

A public water movement came together in Italy to defeat forced water privatization. Article 15 of the 2008 "Ronchi Decree" — named after Andrea Ronchi, Minister of Community Policy from 2008 to 2010 — stipulated that by 2011, water service companies could not be discriminated against by municipal authorities. They were even encouraged to buy up to 70 percent of any listed public water company. Article 154 of the "Environmental Code" stated that the price of water services would be decided on the basis of a guaranteed return on investment. This meant that the private water companies could charge as much as they wanted, to guarantee a higher profit and to further their view of water as an economic good rather than a common good.[7]

The Forum Italiano dei Movimenti per l'Acqua, a network of national associations, trade unions, and local committees opposed to water privatization, launched a campaign to hold a referendum on these laws that drew to a successful conclusion with 1.4 million signatures. The referendum, held in June 2011, was a success. It was the first time since 1995 that a referendum had managed to reach a quorum in Italy. Of the 57 percent of the electorate who participated (about 26 million people), 96 percent voted to keep their water services public.

In Seville, Spain, the Public Water Network was launched in March 2012 — with the motto "Write Water, Read Democracy" — to fight privatization of water services, protect ecosystems, and assert public control over the country's water.

In February 2011, more than 665,000 citizens of Berlin voted in favour of opening the books to disclose details of the 1999 water privatization that had left the city with some of the highest water rates in Europe. Armed with this new information, city authorities ordered the company to cut drinking-water prices. In May 2013, Veolia announced the sale of its shares, giving Berlin Water full public control of water services. The price to the public was punitive, however: 650 million euros — the company's claim for lost profits up to 2028.

Initiative 136 is a powerful citizens' movement in Thessaloniki, Greece, that is fighting privatization and proposing to manage the city's water through local co-operatives. Realizing that what the state has now is not a true public water service, organizers knocked on doors to get Thessaloniki's residents to buy shares in the company and set up a not-for-profit public–community partnership managed by citizen co-operatives. A new national citizens' movement called Save Greek Water was formed in the summer of 2012, and it can claim several victories. That summer, the municipality of Pallini, just north of Athens, adopted a resolution refusing privatization of its water supplies and affirming that they are a commons for the people. In the spring of 2013, the city of Thessaloniki agreed to hold a citizen's referendum on that city's water future.

The European Citizens' Initiative is a citizens' movement determined to hold a referendum on the austerity-backed move to privatize Europe's water services. They want to use a new tool for participatory democracy in Europe, whereby citizens can put an issue on the European political agenda by collecting one million signatures from at least seven different EU member states. Through the referendum process, they will ask the European Commission to enact legislation that implements the rights to water and sanitation as recognized by the United Nations and that promotes the provision of water and sanitation as an essential public service for all. By May 2013 they had collected 1.5 million signatures from eight countries.

In early March 2013, Germany's upper house opposed the EC proposal to promote privatization of communal water supplies. "The Bundesrat attaches great importance to the preservation of the existing structures of municipal responsibility for the drinking water supply," the chamber said in a statement. "The need to ensure a safe, high-quality and health-safe water supply precludes making water a free merchandise."[8]

And in what came as a total surprise, on June 26, 2013, on a recommendation from Commissioner Michel Barnier, the European Commission announced that it had removed water from its concession directive. Drinking water and sanitation services are not to be used as pawns in the escalating conflicts over Europe's austerity program. Barnier gave credit to the citizens' movement to protect water, saying that he understood why people would be angry and upset when told that their water services might be privatized against their will. He added, "I hope this will reassure citizens that the commission listens."[9]

CANADA

The government of Prime Minister Stephen Harper fully backs privatized water. Much like the World Bank, the Harper government will give federal funding to municipal water projects only for improving or investing in new water and wastewater infrastructure that involves private players. Cash-strapped municipalities are now turning to P3s in order to gain access to money.

But not everyone says yes. In November 2011 the citizens of Abbotsford, British Columbia, overwhelmingly voted against a public–private partnership for their water infrastructure expansion, even though the federal government had promised the town $66 million. A growing population and a thirsty industrial base were drying up the water supply in the community, so the city council had come up with a plan to tap into a nearby reservoir and build a filtration and pumping facility and pipeline. The catch was that the contract had to be given to a private company. The mayor of Abbotsford and all the town councillors but one supported the private proposition. They spent $200,000 of taxpayers' money in a marketing campaign to promote a yes vote in a referendum that was also an election for city council.

What city council wasn't ready for was the passionate citizens' movement that sprang up to defend public water in Abbotsford. In the referendum/election, every single councillor who supported the private partnership, including the mayor, was defeated. Patricia Ross, the only councillor to oppose the project, won with an historic number of votes. "We definitely have to roll up our sleeves," she said, "and figure this out — a non-P3 solution."[10]

To counter the Harper government's policy, citizens have launched a campaign to get municipalities to pledge that they will not accept the federal enticement. The Council of Canadians, a large social and environmental justice movement (whose board I chair), and the Canadian Union of Public Employees (CUPE), the largest public-sector union in the country, have joined to promote the Blue Communities Project. The project has adopted a water-commons framework, asserting that water belongs to no one but is the responsibility of all. To qualify as a "blue community," a municipality must adopt three actions: (1) recognize water as a human right; (2) promote publicly financed, owned, and operated water and wastewater treatment services; and (3) ban the sale of bottled water in public facilities and at municipal events.

Already a dozen municipalities in Canada have voted themselves a blue community, and the plan is to expand the movement globally. In September 2013 Bern, Switzerland, became the first non-Canadian city to become a blue community. A number of other Swiss municipalities are showing interest in following suit, as are other public institutions such as schools and hospitals. Many municipalities and universities in Canada see bottled water as an impediment to public water and are banning its sale on their premises. Eighty-five municipalities, fifteen campuses, and eight school boards in Canada have now banned bottled water, and the movement is growing.

This campaign has caught the imagination of young people across Canada who see a way to take direct action to protect public water; it has also been a great teaching tool to address the larger issues. Robyn Hamlyn was a twelve-year-old schoolgirl in Kingston, Ontario, when she saw a film in 2011 about the global water crisis that completely devastated her. She set out to do something. She wrote to the mayor and met with him, then, at his invitation, with Kingston City Council. Moved by Robyn's presentation, Kingston recognized water as a human right.

But Robyn wasn't done yet. With the help of her mother, Joanne, she wrote to dozens of municipalities in Ontario, and by the summer of 2013 she had made presentations to twenty-three of them. Several became blue communities and many adopted new water-protection policies. Shockingly, Robyn became a target of attack by the forces of water privatization. John Challinor, director of corporate affairs for Nestlé Canada, wrote letters to the local newspapers of all the municipalities Robyn had visited and urged them not to listen to her. And an intern with Environment Probe, a pro-privatization research institute, wrote an article for the *Financial Post* chiding local councils for taking their advice "from a 13-year-old."[11]

UNITED STATES

Access to public water in the United States is made more difficult by growing poverty and inequitable wealth distribution. As poverty increases, the problem of water access is exacerbated by rising water rates across the U.S. that are carving a larger bite out of household budgets. In a survey of more than a hundred municipalities between 2002 and 2012, *USA Today* found that water rates had at least doubled in more than a quarter of the locations and even tripled in a few, including Atlanta, San Francisco, and Wilmington, Delaware. The monthly cost of one thousand cubic feet of water in Philadelphia jumped 164 percent in that period, and costs in Baltimore went up 140 percent. Monthly bills top $50 for consumers in many large cities. Noting the need for at least $1 trillion for infrastructure improvements by 2035 to keep up with drinking-water needs, the industry report said that rates would continue to go up.[12]

During those same years, a number of cash-strapped municipalities across the United States decided to go with public–private partnerships in the delivery of water services, which at least partially explains the rising water rates. At present, private companies deliver water services on a for-profit basis to about 15 percent of the American population. But Food and Water Watch reports that there are many problems: maintenance issues in Atlanta, sewage spills in

Milwaukee, corruption in New Orleans, and political meddling in Lexington. To combat this, local communities have become active on this issue and a number of private operations are returning to the public system.

In Felton, California, town residents raised money to buy out American Water, which was a subsidiary of a German transnational. The San Lorenzo Valley Water District purchased the company and now runs the water system for the area on a not-for-profit basis. In the mid-2000s, New Orleans, Louisiana; Atlanta, Georgia; Laredo, Texas; and Stockton, California, all cancelled privatization contracts that were failing their citizens. More recently, after years of protest against high water rates and poor service, residents of Florida were gratified to see the government crack down on the private water utility Aqua America. Unable to make enough profit under the new rules, in September 2012 the company announced it would sell its water systems back to the public water authority.

Food and Water Watch says that these are part of a larger trend to remunicipalize private water operations across the United States, and it is saving money for local communities. In a review of eighteen communities that reclaimed public management of water or sewer services between 2007 and 2010, FWW found that public operations are on average 20 percent cheaper than private operations and that a municipality typically saves twenty-one cents on every dollar by returning its water system to public hands.[13] Food and Water Watch is promoting a federal Clean Water Trust Fund to establish funding to assist communities across the country in keeping their water safe, clean, and public.

Despite these trends, legislation aimed at increasing investment in water infrastructure that favours public–private partnerships was tabled in Congress in 2013. The Water Resources Development Act would offer low-interest loans for water infrastructure projects and would be available to supplement P3s. A second piece of legislation, the Water Infrastructure Now Public-Private Partnership Act, would create a pilot program to explore agreements between the Army Corps of Engineers and private entities as alternatives to public financing. Profits to be made from public funds are huge. A June 2013 report from the Environmental Protection Agency (EPA) found that the nation's drinking-water systems are deteriorating and will need substantial investments if the quality of water services is to be maintained.

Still, grassroots communities are increasingly coming together to fight for both water justice and water conservation, recognizing the need for both if either is to be realized. Not surprisingly, one of the strongest movements started in inner-city Detroit, where so many families have had their water cut off for lack of payment. The People's Water Board was founded in 2009 by a number of groups that included the Michigan Welfare Rights Organization, the Rosa Parks Institute, the Detroit Black Community Food Security Network, the Sierra Club Great Lakes Program, and several labour organizations. The board is made up of representatives of these groups and concerned citizens. They conduct research, hold public meetings, and meet with municipal and state representatives. Their mandate is to achieve access to and affordability of water services and protection and conservation of the water system. They work to ensure that Detroit's water remains in the public trust, free of privatization.

Picketing and protesting one day, writing in-depth analysis another, the group's growing numbers are a political force to be reckoned with. They are rallying to maintain ownership and establish public control of the Detroit Water and Sewerage Department and are keeping the issue of water cut-offs front and centre in the debate. At public events, residents tell of the shame of having their water cut off and the pain of having children removed from the home because of lack of running water. "We focus on the question: what does water mean for all of us?" organizer Charity Hicks told writer/activist Alexa Bradley. "The board has a cross pollinating effect among people focused on poverty, health, growing food, jobs, ecological survival. We attend to both human and ecological sustainability." Hicks warns that authorities need public pressure and that everyone should get involved. "We are saying: you are part of this conversation, you are an expert, we are all experts. We have full agency. This is what democracy looks like."[14]

University of Michigan journalist Lara Zielen sees this as a metaphor for coming struggles about water the world over. "The shutoffs are at the heart of how the Great Lakes are being stewarded. As the world's supply of fresh water dwindles, the Great Lakes will only continue to become more of a focal point. Who gets the water in these lakes and who goes without? The ways in which water equity issues play out in Detroit may foreshadow what's on the horizon for other U.S. cities — and even the world."[15]

It is becoming clearer that the human water crisis and the ecological water crisis are deeply connected and that finding the answer to one will necessitate finding the answer to both. This will require seeing water differently than we do now and putting its protection at the centre of our lives.

WATER HAS RIGHTS TOO

This principle recognizes that water has rights outside of its usefulness to humans and that it belongs to the earth and other species as well as our own. It calls for a new water ethic that puts water at the centre of our lives and protects source water and watersheds in practice and in law. The belief in unlimited growth and our treatment of water as a tool for industrial development have put the world's water in jeopardy. Water is not a resource for our pleasure, profit, and convenience but rather the essential element of a living ecosystem from which all life springs. Since most nation-state laws consider nature and water to be forms of property, it is imperative to create new laws more compatible with the laws of nature. If humans, other species, and the planet are to survive, we must adopt an earth-centred form of governance based on the conservation, protection, and restoration of watersheds and nature. We must build all policies—environmental and economic—around the needs of Mother Earth.

9

THE TROUBLE WITH "MODERN WATER"

What we do to water, we do to ourselves and the ones we love.

— From *Popol Vuh*, the ancient Guatemalan Book of the People

I N HIS BOOK *What Is Water?* Canadian writer, political scientist, and geographer Dr. Jamie Linton presents a challenge to the orthodox way in which we understand water in the modern world. By "modern world" Linton means the post-industrial Western world and its creation of a view of water as a scientific abstraction, removed from its cultural, social, spiritual, and ecological roots. In pre-modern times, Linton writes, water was infinitely varied and diversely known, an aspect of the "history of place," a living element. By reducing water to a chemical compound called H_2O and describing all water, at all scales, as part of a single hydrologic cycle, modern science drove out all the socially and culturally specific qualities of different waters. Now copious supplies of this undifferentiated "colourless, transparent, tasteless compound of oxygen and hydrogen" occurring in a universal form were available to the service of industrial development.[1]

Robbed of its social nature, "entrapped water" became a tool to be measured, quantified, managed, manipulated, removed from watersheds, and controlled for economic growth. The state became an agent in the conquest of water, says Linton, and this "resource" was exploited to yield the highest return. Mighty rivers became hydro power waiting to be harnessed in the service of "hydronationalism" — governments' drive to gain control of their nation's waterways for development. Linton quotes Michael Straus, U.S. president Harry Truman's commissioner of reclamation, who said that control of water was a prerequisite to the kind of development represented by the Hoover Dam and other great American dams built in the 1930s and 1940s.

Modern water drained wetlands and canals, hardened shorelines, and dredged waterways to make way for urban development. Modern water built the Suez Canal and the St. Lawrence Seaway. Its placelessness later allowed it to be shipped around the world in tankers, sold in plastic bottles, and used to grow commodities for the global food trade. Modern water builds river-altering mega-dams and energy-guzzling desalination plants.

Modern water also views groundwater as a legitimate receptacle for our industrial waste. Over the past several decades, U.S. industries have injected more than 120 trillion litres of toxic liquid deep into the earth, using a broad expanse of the nation's geology as an invisible dumping ground, reports ProPublica, an American investigative journalists' consortium. There are more than 680,000 underground waste and injection wells in the United States alone, more than 150,000 of which shoot industrial fluids thousands of feet below the surface. Records show that many of these sites are leaking, although there is no federal oversight of the practice. "In 10 to 100 years, we are going to find out that most of our groundwater is polluted," says Mario Salazar, an engineer who has worked as a technical expert with the EPA's underground injection program. "A lot of people are going to get sick, and a lot of people may die."[2] (Mexico

City plans to draw drinking water from a previously unmapped kilometre-deep aquifer it recently discovered. ProPublica says that this challenges a key tenet of U.S. water policy that water far underground can be intentionally polluted because it will never be used. Had Mexico allowed similar groundwater dumping, the new water find might have been unusable.)

Linton explains that modern water became "imperial water" when it was imposed on the Global South as part of the colonial expansion of the European and British empires. Indigenous relationships with water were often replaced by Western hydrologic technologies. In order to supply the home market in England, for instance, India's water was harnessed and dammed for irrigation, and local knowledge was replaced by the "technological mechanics" of modern water, a process that is still going on.

Modern water viewed arid and semi-arid landscapes as barren and in need of hydrologic re-engineering to become productive. Deserts were greened by massive water diversions that drained lakes and aquifers. Local customs of dryland farming, conservation, and rainwater harvesting were abandoned. Seeing water in this way led to enormous increases in water withdrawals all over the world and allowed water to be transferred to create water wealth in some areas while producing drought in others. Water viewed and used in this way worked to the benefit of the rich and powerful.

The global water crisis is a crisis of modern water. The notion that there is a water crisis rather than a "modern water crisis" has allowed some very powerful interests to take control of the debate. They have defined water as a scarce resource — as opposed to a badly managed one — and this allows them to characterize it as an economic good to be handled in an economical and integrated fashion. Instead of emphasizing the need to reduce water demand and protect local supplies, they seek policies that increase the amount of water available through big infrastructure investment. Defining the water crisis in terms of a world running out of water plays most strongly into the hands of those who would benefit from the provision of water supplies.

Linton goes on to say that private investment in the water supply and improving the economic productivity of water have become the "twin hydrological hopes" of state planners, water experts, and corporate leaders alike. And, of course, the only means to do so is through the market.

BOTSWANA AND ITS MODERN WATER EXPERIMENT

Former Botswana president Festus Mogae is an African superstar. Educated at Oxford and Sussex Universities in Great Britain, Mogae was Botswana's president from 1998 to 2008 and modernized its economy in a way that was pleasing to international market interests. He opened the door to direct foreign investment and privatized essential services. For this and his work to reduce HIV/AIDS in his country, he has received many honours. Former UN Secretary General Kofi Annan praised Mogae for his "outstanding leadership," and he currently serves as a special envoy on climate change for the UN Secretary-General.

Not everyone shares this assessment of Festus Mogae. Survival International, which works to protect the rights of tribal peoples around the world, reminds us that the persecution of the Kalahari Bushmen and expansion of diamond mining in the Kalahari took place under his presidency. Indeed, Mogae once said, "How can you have a Stone Age creature continue to exist in the age of computers? If the Bushmen want to survive, they must change or otherwise like the dodo, they will perish."[3] It is his embrace of all the features of modern water that American journalist James Workman exposed in his book *Heart of Dryness*. However, Botswana is a semi-arid, landlocked country in southern Africa with scarce water resources, but President Mogae wanted a modern state, and a modern state needs modern water.[4]

Botswana's 1991 Water Master Plan (which preceded Mogae) included creation of a system of irrigated industrial agriculture and massive expansion of cattle ranching. For this and for diamond mining — Workman says that De Beers alone has used 11 percent of Botswana's water — the government needed water. To obtain it, they pumped the Kalahari aquifer to feed the Limpopo River. The country punched 21,000 boreholes deep into the desert groundwater, removing hundreds of trillions of litres of ancient fossil water. Eventually the aquifer dried up. Not surprisingly, most of the irrigation water evaporated in the hot African sun, the crops failed, and cattle died.

Mogae added to his country's water woes when he installed modern sanitation systems that use flush toilets, which depend on abundant water sources. The "septic stew" this system created helped extend the distribution of intestinal parasites by providing them with a warm habitat in which to breed. As the population of Botswana's capital, Gaborone, exploded, the sewer system became overwhelmed and spread cholera, dysentery, and diarrhea. As recently as February 2013, the U.S. embassy in Botswana announced that it had tested the water in Gaborone and found it unfit for human consumption: the water coming from residential taps contained waterborne bacteria and biological contaminants.[5]

The Gaborone Dam has supplied the city's water since it was built in the 1960s. But Mogae's intention to create a modern economic and industrial hub in Africa drove up water demand dramatically during his years in power, and pressure on the dam and the Notwane River that supplies it grew exponentially. When Mogae came to power, the dam's reservoir was full; by 2005 it was down to 17 percent capacity. In February 2013 the city's water authority posted a statement that read: "The Water Utilities Corporation would like to inform residents of Gaborone and Greater Gaborone areas that it is currently experiencing water supply challenges. Therefore, they will experience low pressure to no water at all indefinitely."

To Botswana's horror, it turned out that the Gaborone Dam had a grave and potentially fatal flaw: centralized large-scale reservoirs make arid countries increasingly vulnerable to climate change. "After decades of dutifully following foreign advice to the letter," says Workman, "Botswana suffered a midlife crisis. It had copied the best development practices from the United States and Sweden to the UN and World Bank. It had progressed from abject poverty to stable middle-income status. And yet it was now teetering on the verge of a dry nervous breakdown."

LARGE DAMS DESTROY LIVING RIVERS

Botswana's government is not alone in thinking that large dams — the most visible expression of modern water — will tame floods, provide water to the masses, and bring prosperity. The twentieth century saw a frenetic race to harness the rivers of the world. International Rivers, which works to protect rivers and defends communities dependent upon them, says that by the end of the twentieth century the dam industry had choked more than half of the earth's major rivers with some 50,000 large dams. The consequences of this massive engineering program have been devastating.

The World Wildlife Fund reports that large dams (more than 15 metres high) built to provide hydroelectricity and flood irrigation are killing the ecosystems of the major rivers of the world. Only twenty-one (12 percent) of the world's longest rivers run freely from source to sea. The world's large dams have wiped out species, flooded huge areas of wetland, forest, and farmland, and displaced many millions of people. Freshwater environments have the highest proportion of species threatened with extinction,[6] and more than 20 percent of the world's ten thousand freshwater species have already become extinct. Cumulatively, says International Rivers, these dams have re-plumbed rivers in an immense experiment that has left the planet's fresh waters in far worse shape than any other major ecosystem category, even rainforests.

Dams reduce biodiversity, decrease fish populations, lower crop production, disrupt the flow of nutrients needed for water health, and contribute to global warming by trapping methane and rotting vegetation in their reservoirs. Canadian scientists have made a preliminary estimate that reservoirs worldwide release up to 70 million tons of methane and around a billion tons of carbon dioxide every year. Almost one-fifth of India's global warming emissions come from its large dams, says a report by Brazil's National Institute for Space Research.[7]

Big dams also affect the water supply. Toxic algae blooms have rendered some reservoirs unfit to drink. They also cause river deepening, which in turn lowers groundwater. And because they greatly increase the surface area of water, dams increase evaporation. About 170 cubic kilometres of water evaporate from the world's reservoirs every year, more than 7 percent of the amount of fresh water consumed by all human activities. The annual average of 11.2 cubic kilometres of water that evaporates from the Nasser Reservoir, behind the Aswan High Dam in Egypt, is around 10 percent its capacity and is roughly equal to the total withdrawals of water for residential and commercial use throughout Africa.[8]

TURKEY

Turkey, where almost half the population has no access to sanitation, has embarked on one of the biggest, most expensive dam-building projects in the world. There are already 635 large dams in Turkey, but the plans are to build an astonishing 1,700 more by 2023. Virtually every river is to be dammed, and the ancient city of Hasankeyf, thought to be one of the oldest continually inhabited settlements in the world, is to be drowned.

The United Nations notes in a report that the Turkish government has conducted no assessment of the environmental and social impacts of the dams, likely because most of the people affected are marginalized groups: the rural poor, small farmers, nomads, and Kurds. The UN is particularly concerned about the Ilisu Dam on the Tigris River, which will force resettlement of up to seventy thousand people. There is no plan in place for them, something the UN calls "utterly disturbing," and the dam will "severely" restrict the water supply in Iraq, its downstream neighbour. This violates Turkey's "extraterritorial obligations to respect the right to water of the farmers and other residents in Iraq depending on the Tigris River."[9]

The government of Turkey is also planning to sell off rivers and lakes in what Olivier Hoedeman and Orsan Senalp of Corporate Europe Observatory call the biggest privatization of water in the world. The government is privatizing water services, replacing cooperative rural agricultural management with a system of concession rights sold to private companies, and selling off waterways to private interests for periods of up to forty-nine years. "Some countries sell oil. We will sell water," former Turkish president Turgut Özal boasted openly.[10] On a positive note, fierce opposition in their home countries caused the big European banks and companies to pull out of the Turkish dam project and the German, Swiss, and Austrian governments to revoke their export credits guarantees as well.

CHINA

The biggest dam in the world is the Three Gorges Dam in China, begun in 1994. It set records for the number of people displaced (1.4 million), the number of communities flooded (13 cities, 140 towns, and 1,350 villages), and length (more than 600 kilometres). International Rivers says the submerging of hundreds of factories, mines, and waste dumps and the presence of massive industrial centres upstream is creating a festering bog of effluent, silt, industrial pollution, and rubbish in the reservoir.

After years of denial, the government of China recently admitted that there are "urgent" problems with the dam. About 16 million tons of concrete were poured into the barrier across the Yangtze River, creating a reservoir that stretches almost the length of Great Britain and drives twenty-six giant turbines. Algae and pollution that would once have been flushed away plague the reservoir, said the government, and the weight of the water it holds has caused tremors, landslides, and erosion. Meanwhile the Yangtze River, whose delta supports 400 million people and 40 percent of China's economic activity, is experiencing its worst drought in fifty years, damaging crops, threatening wildlife, and stranding thousands of boats and container ships. Regional authorities have declared more than 1,300 lakes in the area "dead."[11]

But admitting the problems plaguing Three Gorges does not mean that dams are losing their appeal to the Chinese government. Half the big dams in the world are in China. Turkey, Iran, China, and Japan account for two-thirds of the big dams currently under construction. Many of China's dams were built hurriedly by the Communist Party in the 1950s and 1960s (proof that modern water is not just a Western phenomenon) and are structurally unsound. Since the 1950s, an average of sixty-five dams a year have burst. The worst involved a breached dam in Henan Province in 1975 that killed

26,000 people. The government admits that 40,000 of its 87,000 dams are at risk of breach, and it has launched a program to repair them. In spite of these failures and that of the Three Gorges Dam, China is ploughing ahead with another hundred large dams planned for the Yangtze alone, and forty-three in the works on the Mekong.[12]

Further, China is building dams around the world — some three hundred of them —from Algeria to Burma. Many of them are projects that the World Bank and Western companies and governments will not touch, reports Denis Gray, Bangkok bureau chief for Associated Press. Poor countries want dams for economic development and to boost living standards. They also see dams as "icons of progress."

China, the world's number one dam builder, is able and willing to finance projects that don't meet international standards, says Ian Baird, a professor of geography at the University of Wisconsin who has worked in Southeast Asia for decades. The consequences, critics say, is a rollback to an era of ill-conceived, destructive mega-dams that many thought had passed. "The most recent trend," writes Gray, "is to dam entire rivers with a cascade of barriers, as China's state-owned Sinohydro has proposed on Colombia's Magdalene River and the Nam Ou in Laos, where contracts for seven dams have been signed." Protests against these dams are growing in Africa, Latin America, and Burma, where China plans to build as many as fifty large dams.[13]

SOUTH AMERICA

Dams are a huge threat to the ecosystem, species, and human populations of La Plata Basin, the second largest river basin in South America, which crosses Paraguay, Brazil, Argentina, Uruguay, and Bolivia. The basin is home to a vast array of wildlife, including 650 species of birds, 260 species of fish, 90 species of reptiles, and more than 80 species of mammals, including ocelots, jaguars, and tapirs, according to the World Wildlife Fund. It is the site of the largest freshwater wetland in the world.

But it is also home to the Itaipu Dam, the largest power plant on earth, capable of providing more power than ten nuclear stations. Its construction dammed the Paraná River, the second biggest river in South America, and flooded 100,000 hectares of land. In doing so, it drowned Guaíra Falls, the world's largest waterfall by volume, and a national park. The basin faces the largest number of planned dams in the world after China, says the w w f, twenty-seven of them. This will cause massive containment of river waters and destroy the headwaters of several rivers.

The biggest of these projects is Hidrovia, a plan by the five countries the rivers pass through to convert the Paraguay and Paraná Rivers into an industrial shipping channel. Backed by the Inter-American Development Bank and the United Nations Development Programme (both friends of modern water), the original plan would have dredged and redirected the rivers to create a 3,442-kilometre-long navigation channel that would allow cargo ships access to the interior of the continent during the dry season. The World Wildlife Fund warns that Hidrovia would seriously exacerbate the loss of water inflow caused by climate change, increase downstream flooding, and displace many indigenous communities. While fierce resistance has managed to stop Hidrovia for now, the demands for it and for other dams in the region are still strong.

Indigenous people and peasants have successfully slowed construction of the Belo Monte Dam, on the Xingu River in the state of Pará, Brazil. If it is built, Belo Monte will be the third largest dam in the world and will cover 500 square kilometres of tropical forest, flooding part of the territory of the Kayapo people. In the dry season an additional 6,000-plus square kilometres will be flooded, affecting

more than fifty thousand people. In March 2013 the Brazilian government ordered troops to the site to control the protests and
confrontations over the dam, which are expected to grow.

DESALINATION ADDS TO THE WATER CRISIS

In spite of its high production costs, desalination as an answer to
water scarcity is growing. From 2001 to 2011, the industrial capacity
for desalinated water expanded 276 percent, to 6.7 billion cubic metres
a day, according to the International Desalination Association. More
than sixteen thousand plants are now in operation and the industry
is growing by 15 percent a year.[14] While the Middle East currently
claims the highest number of desalination plants, China will soon
surpass it, more than tripling its production in the next few years.
Desalted water will supply 15 percent of the needs of China's factories along its industrial eastern seaboard, according to the Chinese
National Development and Reform Commission.

There are two commercial technologies for water desalination,
both energy intensive. The first, and most common, is to force seawater through chemical-laced membrane filters at high pressure. The
second uses condensation and evaporation. Aside from using large
amounts of energy, these technologies dump concentrated salt back
into the oceans, harming local wildlife and polluting the seawater.
In its report *Desalination: An Ocean of Problems*, Food and Water
Watch found that desalination takes nine times as much energy as
surface-water treatment and fourteen times as much as groundwater
production.

As well, desalination creates nearly twice the emissions as treating and reusing the same amount of fresh water. Intake structures
kill billions of fish every year in the United States alone. Desalinated
water costs twice as much as tap water, not counting the cost of the

energy to produce it. Desalination invites corporate control of the water supply, said the report, and warned that private corporations plan to sell desalted water to the public at a premium.[15]

Studies show that water conservation and infrastructure upgrading would save more water than desalination could ever produce, and much more cheaply. Peter Gleick, of the California-based Pacific Institute, has shown how California could save a full third of its current water use through conservation, 85 percent of which could be saved at a cost lower than that of producing new sources of water from desalination.

Bloomberg News gives a stark example of the contradiction inherent in seeing water as a commodity, in a March 2013 report on the water crisis in Latin America. Reporter Michael Smith describes how the residents of the city of Copiapó, Chile, suffer daily tap-water cut-offs as foreign mining companies drain local aquifers for their mines. They are operating in the Atacama Desert, home to the biggest copper mines in the world, and an area so dry that rainfall has never been recorded in some places. Instead of cutting off the mining companies and seeing to the needs of the people, the government of President Sebastián Piñera convinced several mining companies, including London-based Anglo American, to build desalination plants 60 kilometres away on the Pacific Ocean and pump the desalted water to their mines.[16] While the company claims it is saving the water supplies of the region, it is using large amounts of energy to produce and ship this desalted water. In doing so, it is creating more greenhouse gas emissions in one of the driest areas in South America.

RUNNING OUT IN THE MIDDLE EAST

Desalination is seen by many as the answer to the water crisis in the Middle East. This has given some countries the false impression that they can use as much of their scarce water supplies as they like to build cities and irrigate deserts. Seventy percent of the world's desalination plants are in this region, mostly in Saudi Arabia, the United Arab Emirates, Kuwait, Bahrain, and Israel. These plants are pouring so much concentrated salt back into the Arabian Gulf that experts say they need to start talking about "peak salt" — the point at which the Gulf becomes so salty that relying on it for fresh water stops being economically feasible.

The Abu Dhabi newspaper *The National* quotes Dr. Shawki Barghouti, director-general of the International Centre for Biosaline Agriculture in Dubai, who is deeply concerned about how much brine is being dumped back into the ocean. He says that the damming of so many rivers in the region has cut the flow of fresh water into the Gulf, water that would have diluted the salt concentrations. In addition, the water that does flow into the Gulf is polluted. In the north, the outflow of the Shatt al-Arab River carries with it the polluted waters of Basra and Baghdad, and many countries, including Kuwait, Saudi Arabia, and Qatar have substandard water-treatment plants. This pollution is made worse by the thousands of tankers entering the Gulf that wash their tanks illegally. All of this is exacerbated by the Gulf's small size, relative shallowness, and slow circulation. "Between the tankers, pollution from urban centres, and the brine disposed from desalination plants, the Gulf is almost dead," says Barghouti.[17]

Israel is another water-challenged country that is increasingly dependent on desalination for its water needs. Its aquifers are drying up from overuse and the Sea of Galilee, known in Israel as Lake Kinneret, has fallen to record low levels. The country has three large desalination plants in operation, one of which is the largest reverse-osmosis water plant in the world. Two more plants are to be operational by 2014. When they are completed, desalinated water will supply 50 percent of Israel's drinking water. Many environmentalists are deeply concerned about the red plumes of iron concentrates these plants spew into the already heavily polluted Mediterranean. The United Nations Environment Programme has warned that dumping of sewage, mercury, phosphates, and other pollutants has put most of the Mediterranean's marine life in danger. The pollution from these desalination plants adds to the burden placed on these waters.

OIL WEALTH HIDES THE CRISIS

The Water Project, a U.S.-based charity, notes that the combination of large oil reserves and scarce water supplies has created a particularly deadly combination of factors in the Middle East. Desertification is a sweeping environmental problem in countries such as Syria, Jordan, Iraq, and Iran. The greatest culprits are unsustainable agricultural practices and overgrazing. Agriculture uses 85 percent of the water in the Middle East, and dams and diversions for heavy irrigation are destroying water sources at an alarming rate.[18]

The Dead Sea borders Israel, Jordan, and the West Bank and is the lowest terrestrial point on the planet. It is dying. Diversions to the Jordan River have led to water levels falling more than one metre a year, and a drop in the flow of the Jordan River of 98 percent. Thousands of sinkholes have opened up around the Dead Sea's coastal plain, created when groundwater from adjacent aquifers flows in to replace retreating seawater.[19]

Scientists fear that time is running out for a solution in another water-stressed country, Yemen, which can no longer grow food to sustain its people. Its streams and natural aquifers are becoming shallower every day, and Sana'a risks becoming the first capital city in the world to run dry. Unregulated drilling has caused the water table to drop by 1,200 metres in some places. More than half of this water goes to cultivating khat, a narcotic plant that feeds no one.[20]

The problem is region-wide. A disturbing new study using data from a pair of gravity-measuring NASA satellites found that large parts of the Middle Eastern region have lost far more freshwater reserves in the past decade than previously thought. In the seven-year period from 2003 to 2010, parts of Turkey, Syria, Iraq, and Iran along the Tigris and Euphrates river basins lost 144 cubic kilometres of total stored fresh water. That is almost the amount of water in the Dead Sea. The researchers attribute about 60 percent of the loss to the pumping of groundwater, and principal investigator Jay Famiglietti has warned that the area cannot sustain this kind of water loss.[21]

Great oil wealth in some countries has given the region the false impression that it can buy its way out of the crisis. The wealthier Arab states, primarily the oil producers of the Persian Gulf, have no rivers and little rain. The Arab world has 5 percent of the world's population but only 1.4 percent of the planet's renewable freshwater supply. By 2025 it is estimated that the Arab population will total around 568 million, gravely stretching the region's shrinking water resources. The World Bank says that climate change will result in a 25 percent decrease in precipitation and a matching increase in evaporation rates by the end of this century. A 2010 report by the Arab Forum for Environment and Development said that by 2015, Arabs will have to get by on less than one-tenth of the world's average per capita water allocation.[22]

The United Arab Emirates is especially vulnerable. A report from the UAE Industrial Bank notes that the Emirates has the highest per capita consumption of water in the world. At its current rate of water withdrawals, the region will deplete its natural freshwater reserves in fifty years. Yet to look at the spectacular urban growth in the area, one would never know there is a problem. The Water Project notes that the UAE is famous for its luxurious cities filled with lavish resorts, shopping malls, and gleaming highrises.

Few places in the world can rival the excesses of the UAE city-state of Dubai. Thirty years ago Dubai was desert, inhabited by tumbleweed and scorpions. But today, thanks to oil money, it is a global testament to urban extravagance, boasting the highest skyscraper in the world, some of the world's largest and most expensive shopping malls, a ski slope with real snow, and the Palm Islands, artificial islands built in the shape of palm trees that house major luxury hotels, resorts, and golf clubs.

One Palm Island houses the famous Atlantis Hotel, an underwater-themed resort that cost $1.5 billion to build. The Atlantis boasts 1,500 guest suites, two three-storey underground suites with views directly into the Ambassador Lagoon, and a 17-hectare "Aquaventure Waterpark" that includes a 2.3-kilometre river slide, complete with cascades and tidal waves and a Mesopotamian-style ziggurat temple reaching more than 30 metres into the sky that features seven water slides — two of which catapult riders through shark-filled lagoons. The hotel also boasts the Lost Chambers Aquarium, an underwater exhibit with more than 65,000 fish and sea creatures, and the Ambassador Lagoon, an 11-million-litre marine habitat with a viewing panel overlooking the mythical ruins of Atlantis.

In his report "The Dark Side of Dubai," British journalist Johann Hari tells of the human and environmental cost of such extravagance. Thousands of foreign workers are housed in a vast concrete wasteland an hour out of town; they are bused in every day to work fourteen hours in the desert heat for low wages. Construction companies confiscate their passports when the workers arrive and many are treated like slaves, says Hari, who recounts many harrowing examples of human rights abuses.

But the ecological cost is equally devastating. Hari speaks of standing on one of Dubai's many manicured lawns and seeing water sprinklers everywhere. He reports that the Tiger Woods Golf Course would need 16 million litres of water every day, and that the heat is so intense it cooks everything that is not artificially kept constantly wet. The water comes from the sea through desalination plants, making it the most expensive water on earth. The energy needed to supply this mirage is the reason why the residents of Dubai have the biggest average carbon footprint on earth.

Hari reports on a conversation with Dr. Mohammed Raouf, the environmental director of the Gulf Research Centre, who warns that depending on this expensive water — and so much of it — will make Dubai vulnerable in an economic downturn. If the world shifts to an alternative to oil and the area loses its financial clout, it would be a catastrophe. Dubai has enough water to last only a week.[23]

The laws of nature have been broken in Dubai and Botswana, in Las Vegas and in fact all over the world. Water has become a tool for a certain vision of modern life and for the human elite the world over. The loss of respect for water is entering a lethal new stage in the global search for resources and the demands of a global market.

10

CORPORATE CONTROL OF FARMING IS EXTINGUISHING WATER

The water wars that the popular media would have us believe to
be inevitable will not be fought in the battlefield between opposing armies, but on the trading floors of the world grain markets
between virtual water warriors in the form of commodity traders.
— Anthony Turton, South African scientist and resource manager

THE WORLDWATCH INSTITUTE HAS a warning for the world:
our method of food production is using and fouling far too much
water and creating large amounts of greenhouse gases. Governments
and transnational food and seed companies promote a system of
global trade that pits farmers in one country against those in another
to keep the system "competitive." This allows for vertical control of
commodities by the big players. Corporate-controlled food production is a classic case of modern water.

This system squeezes out small-scale farmers and has a negative
impact on the biodiversity needed for healthy watersheds. Industrial
agriculture accounts for $1 trillion of the economy — which is why
corporations and governments are so drawn to it — but also at least

70 percent of water withdrawals and 15 percent of greenhouse gas emissions. Governments reward farmers for sheer quantity, with little guidance on the environmental impact of their operations.

Big agribusiness also overtaxes rivers with the water they draw for irrigation, and many now dry up before they reach the sea. These include the Yellow River in China; the Amu Darya, Syr Darya, and Indus in Asia; the Euphrates and the Tigris in the Middle East; the Colorado and Rio Grande in North America; and the Murray and Darling Rivers in Australia, all of which have been over-tapped for industrial agriculture. To avert ecological disaster, Worldwatch calls for a change in the way we farm, away from large-scale industrial farming to small-scale local, community-based food production.[1]

DUST BOWL LESSONS FORGOTTEN

Some have ranked the North American Dust Bowl of the 1930s as one of the three greatest ecological disasters in history. It was a six-year period of severe dust storms caused by prolonged drought, over-cultivation of marginally productive land, and poor farming methods. Extensive clearing of the Great Plains of Canada and the United States for agriculture displaced the natural prairie grasses that held the soil in place. The soil dried up and literally blew away, causing the severe and frequent dust storms. In May 1934 the greatest dust storm ever recorded blew more than 300 million tons of topsoil across the continent, degrading more than 7 million hectares of agricultural land.[2] Forty-five hundred people died of the heat before the drought ended.

It is traditional wisdom that the drought was an act of God, an unpredictable event that wreaked havoc on soil badly managed. But scientists are beginning to understand that destruction of vegetation can alter rain patterns and can in itself cause drought. When the water cycle is disrupted because the vegetation has been removed, water vapour is lost to the local watershed. Studies now show, for instance, that cutting down rainforests actually reduces the amount of precipitation in the area. New research from the Earth Institute at Columbia University reports that the Dust Bowl dust storms amplified the natural drop in rainfall and turned an ordinary dry cycle into a natural and agricultural disaster. The dust doubled the loss of rainfall and even moved the drought further north, into other farming regions.[3]

In his provocative book *Restoring the Flow*, Canadian water researcher and author Robert Sandford writes about the lessons learned and forgotten about the Dust Bowl, lessons we would be well advised to relearn. One is that people who were profiting from afar — the non-resident operators who comprised one-third of farmers in the 1930s — abandoned the land first, having no interest in it over the long haul. Another was that the Dust Bowl was caused by ecological breakdown and land misuse, and if the problem was not to repeat itself, attitudes and practices would have to change dramatically.

In December 1936, reports Sandford, advisors from the U.S. Bureau of Reclamation, the Works Progress Administration, and the National Resources Committee placed a 194-page report called *The Future of the Great Plains* in front of President Franklin D. Roosevelt. They blamed the Dust Bowl on "attitudes of mind" that allowed an expansionist free-enterprise culture to exploit the Great Plains, and they proposed a new model for regional land-use planning that is still relevant, says Sandford.

Several destructive self-deceptions led to the crisis: the domination-of-nature ethic, which reduced the land to nothing more than a raw material to exploit; the view that natural resources were inexhaustible; the notion that what is good for the individual is good for everybody; the idea that a property owner can do whatever he likes with his property; the faith that markets will expand indefinitely; and the notion that factory farming is desirable. This led to irresponsible non-resident ownership, speculative commercialism, and land-abusing tenancy.

As a result of this report and its detailed recommendations, a program to restore the land and water of the Great Plains took shape and an extensive long-term program of soil conservation came into existence. This included a commitment to ending factory farming; tight controls over agricultural investments, profit making, and land ownership; and establishment of a permanent and enduring Great Plains culture. For a time these values, practices, and conservation measures worked.

But Sandford points out that memories are short. When the rains came and the profits returned, the lessons were forgotten. With the Second World War came expanded opportunities for export and the rise of large-scale corporate agricultural interests. Business farming neglected the precepts of limiting land and water use to protect the environment for the long haul, and agri-monoculture was established on the Great Plains. "Six years of Dust Bowl life was apparently not enough to cause a change in the fundamental societal values that favoured the endless pursuit of increased wealth over the health of the Great Plains ecosystem,"[4] says Sandford. While conservation programs and set-asides remain in place, government subsidies to farmers for conservation have fallen to an all-time low in the past decade. This money simply cannot compete with the profits to be had from the ethanol boom, which makes it financially attractive to use targeted lands for production, despite conservation incentives.[5]

FOOD CORPORATIONS CONTROL THE SYSTEM IN THE UNITED STATES

In her powerful new book *Foodopoly*, Food and Water Watch executive director Wenonah Hauter says the retreat from sustainable farming was not an accident. Rather, it was the result of farm and food policies that were first proposed by some of the most powerful interests in business and government after the Second World War. These men envisioned a future in which most young rural men would supply labour for manufacturing in the industrial north rather than continue farming, and food would be grown on a small number of large industrial farms. They foresaw a future in which food production would be globalized for economic efficiency, and the "free market" would create the cheap inputs necessary for processed food. They mapped out a postwar program to expand chemical-intensive agriculture and to grant industrial and financial interests over it. Farmers became the target of policies aimed at reducing their numbers, says Hauter, policies that later became enshrined in law.

As a result, millions of American family farms have folded in the past several decades as government policy encourages larger, more intensive farm operations such as factory farming. A handful of seed, meat, and dairy corporations now dominate most aspects of the food system, giving them enormous control over markets and pricing and allowing them to influence farm policy. Hauter is deeply critical of corporate control of the food chain. "Big business thinks of our kitchens and stomachs as profit centres," she writes. Food and agricultural products have been reduced to a form of currency on income statements. The worth of these products is measured in return on investment or opportunities for mergers or acquisitions. Their value is described in a "Wall Street–speak of deals, synergies, diversification, and 'blockbuster game changers.'" The results have been negative for all but the big companies. Consumers are paying to maintain the corporate profits taken at each stage of production.[6]

And unbridled industrial expansion on the Great Plains, combined with climate change, may be affecting rain patterns and causing drought to this day, say Benjamin Cook and Richard Seager, the authors of the Columbia Earth Institute study. Canadian water scientist David Schindler further explains that part of the effect of devegetation is to increase the speed of runoff. When it does rain, there is no vegetation to intercept the water. Dried-up, compromised wetlands also hold less water. So not only does it rain less, less of what fall stays. And with rising temperatures as the climate warms, evaporation steals more and more of what falls.

In fact, the American West is heating up nearly twice as fast as the rest of the world, reports the Rocky Mountain Climate Organization and the Natural Resources Defense Council. In the five-year period from 2003 to 2007, the average temperature in the Colorado River Basin, which stretches from Wyoming to Mexico, was 2.2 degrees Fahrenheit hotter than the historical average for the twentieth century. That rise was more than twice the global average increase of 1.0 degree during the same period.[7] There is a clear connection between current farming methods and land use and the increasing drought and warming in the U.S. West.

As well, over-pumping of the region's groundwater is leading to disaster, say scientists. In 1953 the centre-pivot irrigator was patented, unleashing an assault on the West's groundwater. This method of crop irrigation uses pipes that rotate on a pivot to relentlessly spray pumped groundwater, creating "crop circles" that can be seen from the air. A May 2013 story in the *New York Times* says the centre-pivot irrigator helped start a revolution that raised farming from hard-scrabble work to a profitable business and led to exponential growth in water-intensive production of food such as corn. Since the pivot's debut, in Kansas alone the amount of irrigated cropland has grown more than ninefold. The *Times* calls the pivot a "villain" that has led to drying up of the High Plains Aquifer, including the Ogallala Aquifer.[8]

In 2010 the U.S. Department of Agriculture reported that the Ogallala Aquifer would become largely depleted in our lifetime. Oregon environmental writer William Ashworth, author of the book *Ogallala Blue,* says that if the aquifer goes dry, more than $20 billion worth of food and other fibre will disappear from the world's markets. "Groundwater mining," he said, "is not an accident here. It is a way of life. It is also a way of death."[9]

Corporate farming is deeply harmful to water and watersheds. Four companies, led by Tyson and Cargill, control 83 percent of the beef-packing industry. Small and medium-sized farms have been replaced by factory and industrial farms that confine thousands of cows, hogs, and chickens in tightly packed warehouses, where they generate tons of liquid and solid waste. Hauter says the world's largest beef producer, Brazil's JBS, owns the Five Rivers Cattle Feeding company, which has a capacity of 980,000 head on thirteen feedlots in several U.S. states, as well as Canada (Brooks, Alberta). The largest, in Yuma, Colorado, alone has a capacity of 110,000 head.

The feces and urine produced by the hog farms of corporations such as Smithfield, the largest pork producer and packer in the world, contain ammonia, methane, hydrogen sulphide, cyanide, phosphates, nitrates, heavy metals, antibiotics, and other drugs. They fall through catchment floors to lagoons below that can hold as much as 180 million litres of toxic waste water. North Carolina's 11 million hogs create massive amounts of sewage. In just one of these factories, 2,500 pigs produce 100 million litres of liquid waste, 4 million litres of sludge, and 44 million litres of slurry (water plus manure) per year. Over the past decade, many lagoons have spilled their toxic contents into local waterways, and hog waste is largely considered the culprit for a massive fish kill in the Neuse River in 2003.

Slaughterhouses also badly pollute local water systems. Hauter

writes about the Smithfield slaughterhouse in Tar Heel, North
Carolina, the second biggest in the world, which slaughters some
34,000 hogs every day. It pulls 8 million litres of water daily from the
local aquifer and returns about 12 million litres of waste water to the
Cape Fear River.

THE ALGAL BOWL

The runoff from factory farming is damaging waterways. Nutrient
overloading from fertilizers and human and animal waste causes
eutrophication and growth of blue-green algae. Nitrogen and
phosphorus, by-products of intensive livestock operations, cause
water-based plant life to multiply. When they reach a certain density,
they can kill a lake or stream. Renowned Canadian water scientist
Dr. David Schindler, of the University of Alberta, writes in his seminal book *The Algal Bowl* that, whereas the Dust Bowl arose from
mismanagement of land, the Algal Bowl arises from mismanagement of both land and water, because the mismanaged land generally
lies in the catchment of a lake or stream.[10]

Schindler quotes the late John R. Vallentyne, a freshwater scientist, formerly with the Canadian Department of Fisheries and Oceans,
who wrote the first edition of the book back in 1974. In it he predicted
that North Americans would be living in the middle of an Algal Bowl
by the year 2000, with degradation of water comparable to that of
the land during the Dust Bowl. Vallentyne's predictions have largely
come true, says Schindler, partly for reasons he anticipated and partly
because of climate change and the rapid expansion of livestock culture, which were not foreseen at the time. Once more prevalent in
Europe and eastern North America, eutrophication has been some-
what reduced through strong laws in Europe, a concerted effort to deal
with its presence in the Great Lakes, and the phasing out of phosphates

in detergents. But the growth of factory farms and laxer regulations in western North America has moved the problem west. The Algal Bowl now threatens the region that was once home to the Dust Bowl.

The global food trade has created a huge market for meat, and much of North America's production is shipped to the Far East, where demand is growing. Intensive livestock operations have exploded as a result. According to the United Nations Food and Agriculture Organization, (FAO), if we include land used to grow crops to feed the animals, 70 percent of agricultural land is now used for livestock production. In the United States, 70 percent of the grain that is grown feeds farm animals, and more than half of the water used daily for all purposes goes to livestock production.

Forests, wetlands, and riparian areas have been cleared for factory farms, and the rules for disposal of waste are in many places, including North America, "primitive." An average beef cow produces eleven times more phosphorus in its waste than a human, and an average hog produces ten times more. A medium-sized factory farm can easily produce as much phosphorus as a mid-sized city, but without the sewage disposal of the city. Where factory farms are concentrated, intense damage can be done to a single watershed.

Runoff from feedlots in Texas is largely responsible for a 12,000-square-kilometre dead zone in the Gulf of Mexico, one of the largest in the world. Dead zones are low-oxygen areas of the world's oceans and large lakes caused by nutrient loading and eutrophication. Intensive hog farming is largely blamed for the state of Canada's Lake Winnipeg, the tenth largest lake in the world and often referred to as "Canada's sickest lake." "What was once a small patch of algae...now grows to smother more than half of the 24,500-square-kilometre lake most summers," reports *Maclean's* magazine.[11] The "putrid green mat," twice the size of Prince Edward Island, is killing the fish and destroying the tourist industry. Some scientists say Lake Winnipeg is already dead.

The problem is growing everywhere. Water-rich New Zealand has put its lakes, rivers, and groundwater at risk for its dairy farming industry, which now supplies much of Asia with powdered milk. Forests and wetlands have been cleared to make way for massive dairy herds right across the island. The industry, which uses pivot irrigation, is depleting groundwater sources; each of the country's 6.5 million cows requires about ten thousand litres of water a day. Nitrate waste is also poisoning the watersheds. The New Zealand Green Party says that more than half the rivers and a third of the country's lakes are unsafe for swimming, and the groundwater in some areas has nitrate levels that are ten to twenty times above natural levels. Two-thirds of the native freshwater fish are at risk or threatened with extinction. Blue-green algae are now widespread, and in 2013 New Zealand experienced its worst drought in seventy years.

The International Lake Environment Committee Foundation (ILEC) of the United Nations Environment Programme undertook a survey of the state of the world's lakes, examining the 217 largest lakes for a variety of problems. The ILEC found that eutrophication has grown in every one of the lakes over the past fifty years. Lake restoration in some wealthy countries has stemmed eutrophication but not solved the problem, a result of the amount of sediment already stored in the lakes. The most successful lakes were in areas with little agriculture.

The problem is reaching crisis proportions in developing countries, said the UN, because of the expense of pollution abatement. Lakes Dianchi and Taihu, in China, are covered by dense algae blooms; fish breeding has been abandoned, as there is no oxygen in the water. Dense plant mats cover parts of Lake Victoria in Africa, and many fish species have become extinct. The ILEC reports that even Siberia's Lake Baikal, the largest freshwater body in the world and 1.7 kilometres deep, shows signs of eutrophication, including decreased transparency and increased concentrations of algae. "A solution to eutrophication in the developing countries is urgent," says the report, "since stopping eutrophication becomes more and more difficult and expensive every year it is postponed due to increasing nutrient accumulation in sediments."[12]

The solution to eutrophication is simple, says David Schindler. We must prevent excess nutrients from entering lakes; keep the inflows, outflows, and wetlands in the catchments of lakes intact; and allow the food chains and fish habitats of lakes to remain in their natural states.

The problem is political. The world's leaders, who promote the global food trade, are deaf to the science of overuse, despite the fact that the problem is intensifying.

TRADE IN VIRTUAL WATER THE HIDDEN STORY

Another threat to the world's water arises from our system of food production. The global trade in food is really about the trade in water. It takes a great deal of water to grow food: 140 litres for a cup of coffee and 2,400 litres for a hamburger. The water used to produce food is called "virtual water," and when the food is exported, the water embedded in it is exported too, right out of the watershed and the country. This is linked to our "water footprint" — all the water required for human activity. Increasingly, in a world of global food markets, the question is whether a nation-state is supplying its water footprint from its own sources or whether it is depending on the water sources of other countries and regions to supply its population with its water needs.

The concept was developed in the early 1990s by J. A. Allan, a geography professor at King's College, London, to indicate the amount of water made available through the agricultural commodity trade. Arjen Y. Hoekstra, a scientist at the University of Twente, in the Netherlands, and scientific director of the Water Footprint Network, further refined the concept by including the water used at all stages of the commodity process, including waste water, as making up the "virtual water content" of a commodity. The virtual water content consists of all the different components:

- blue water — the water in aquifers, rivers, lakes and runoff;

- green water — water from rain directly transpired by vegetation;

- brown water — water contained in soil; and

- grey water — wastewater from the process of production.

As an aside, there is virtual water in most services and products as well, from energy to clothing. For instance, it takes 2,400 litres of water to produce a cotton shirt. It also takes a lot of water to manufacture a computer chip, and even more to run the enormous computer "server farms," which are clusters of network data servers housed in giant warehouses. An Amazon.com data centre manager has estimated that a fifteen-megawatt data centre can use up to 1.5 million litres of water a day to cool the computers.[13]

Until recently our per capita water use was calculated as being the water we physically handle and consume: the water we use for cooking, gardening, bathing, and so on. The number varies widely from region to region. But a comprehensive 2012 study carried out by Arjen Hoekstra and his colleagues says that if we include all the water embedded in the food we eat and the other products we consume, the global per capita footprint is four thousand litres of water per day — ten to twelve times higher than has previously been estimated. The study, one of the largest ever done on this subject, quantified and mapped the global water footprint. It found that water used by the agricultural sector now accounts for almost 92 percent of annual freshwater consumption (as opposed to the 70 percent generally cited by the UN, the World Bank, and others). The increase, say the authors, is due to the growth in the global food trade.[14] These are shocking numbers that demand an examination of early assumptions about the global trade in virtual water.

When Allan first envisioned the concept of virtual water, he saw it as an instrument through which water-scarce states could conserve domestic water by importing from water-rich states and thereby achieve water security. He also promoted the notion of a virtual water trade as a way of achieving efficiency in global water use. While Hoekstra reports that international trade does reduce global water use in agriculture by a very small amount (5 percent) — because water-intensive commodities are traded more often from countries with high water productivity to countries with low water productivity — current evidence suggests that there is no rational planning behind the food and virtual water trade, just competition driven by market forces.

According to a study published by Vijay Kumar and Sharad Jain in *Current Science*, "Analysis of country-level data on fresh-water availability and net virtual water trade of 146 nations showed that a country's virtual water trade is not determined by its water situation." In fact, say the authors, virtual water is exported out of countries that are rich in land but not necessarily rich in water.[15] One major factor to understand is that farming is a consumptive use of water; that is, the water is not returned to its original source. As explained in a report on virtual water for the Council of Canadians by researcher Nabeela Rahman and Blue Planet Project campaigner Meera Karunananthan, water-intensive production for domestic consumption does not have the same impact as water-intensive production for export. Once exported, the water embedded in the product is removed entirely from the local watershed.

In a world with scarce water resources, the ability to import

water-intensive products reflects an invisible source of power for
states that are able to conserve their own water resources by importing water-intensive products.[16] Germany, for example, is not a
water-scarce country but is a net importer of virtual water. It leaves
its water footprint in countries such as Brazil, the Ivory Coast, and
India, from which it imports coffee, cotton, and other goods. The
United Kingdom imports two-thirds of its water footprint. Saudi
Arabia and Japan import most of their water footprint.

Wealthy countries are able to maintain their water security by
relying on other countries for water-intensive products; they see
virtual water as an alternative to their own sources of water. The
Hoekstra study highlighted how patterns in international commerce
create disparities in water use. For the first time we have a spatial
analysis of water consumption and pollution based on worldwide
trade indicators. The study linked the world's water footprint to
free trade and found that more than one-fifth of the world's water
supplies go towards crops and commodities produced for export.
Hoekstra anticipates a "drastic" change in consumption in China as
it relies increasingly on farmland in Africa. This, he says, will lead to
greater water imports.

"These are all clear indicators," noted a University of Twente
press release on the Hoekstra study, "that water scarcity is not
a local problem but must be seen from a global perspective. The
researchers are therefore questioning whether the continued use
of the limited blue WF (water footprint) for export is a sustainable
and efficient option." Rahman and Karunananthan point out that,
unlike the bulk water exports that generally fuel public outrage,
virtual water exports are a more covert form that helps national
leaders avoid fuelling the political unrest that would come with
public awareness. Or, as J. A. Allan said, "It prevents water crises
from becoming water wars."[17]

But the damage is real. When water is removed from a watershed, it is removed from the local hydrologic cycle as well. That in turn reduces evaporation, heating up the atmosphere and bringing climate chaos. It is no coincidence that deserts are growing in more than a hundred countries. And the damage is not confined to poor countries. The United States, Canada, Brazil, and Australia export virtual water as well.

Americans, who lead the world in virtual water exports, are stunned to learn that they export about one-third of their daily water withdrawals in commodity exports. Most of this water comes from the states with the least water. California exports large amounts of water-intensive hay and more than half its rice production — another water-intensive crop — to Japan. British environmental writer Fred Pearce says that the virtual water trade is emptying the Colorado and Rio Grande Rivers and sucking the Ogallala Aquifer dry.[18]

Australia, the driest inhabited continent on earth, is the world's largest net virtual water exporter, meaning that it sends more water out of the country in virtual form than it imports. For decades the Murray–Darling River system has been exploited for irrigation to produce rice, cotton, wine, and other water-intensive products for the world market. Cotton production alone uses 40 percent of the total water extracted for irrigation from the Murray–Darling, and the majority of that cotton — and the water — is exported. Australia is a net exporter of just under 64 billion cubic metres of virtual water each year, shipping out much more water than it takes in.

Dr. Ian Douglas of Fair Water Use Australia says, "While few would question the important contribution made by most Australian farmers, it defies belief that a country continually struggling with unreliable rainfall and severe drought allows more virtual water to be lost than any other nation on the planet." Douglas says the situation is made even sadder by the fact that the government is having a hard time finding a fraction of this amount to return to the river for restoration.[19]

Even Canada, with its supposed water abundance, is putting its water supplies at risk with its relentless increase in commodity exports. Canada is a net virtual water exporter, second only to Australia. Canada's net annual virtual water exports would fill the Rogers Centre in Toronto 37,500 times. Every year, Canada exports virtual water stored in wheat, barley, rye, and oats that is equivalent to twice the annual discharge of the Athabasca River. These virtual water exports are placing a huge burden on the province of Alberta. With just 2 percent of Canada's water supply, Alberta accounts for two-thirds of the water used for irrigation, much of it for export.

University of Alberta scientist David Schindler expects Alberta to become the first water "have-not" province in Canada. Its water supply is already fully allocated, but ambitious free-trade plans of the Canadian and Alberta governments have created a huge increase in intensive livestock operations in anticipation of large export demand. Livestock water use in Alberta is set to double in the next decade. No one knows where that water will come from.[20] Ironically, in most of these examples governments allow agribusiness unlimited access to free or cheap water, a clear case of subsidizing that increases their bottom line and a transfer of the water commons into private hands.

Our global system of corporate-dominated, market-driven food production is using an unacceptable and unsustainable amount of the world's fresh water. Water is plundered from watersheds, rivers, and aquifers and shipped all over the world. Waterways are fouled and dying in the name of trade and profit. In this system, water has not been factored in as a non-renewable cost of production, and its unlimited use supports a giant food market that knows no boundaries. As demand grows, the capacity of watersheds to absorb this abuse decreases while the demand remains relentless. The UN estimates that by 2025 we will have to earmark an additional two thousand cubic kilometres of water for irrigation to meet the demands of the global food trade. This is equivalent to almost a billion Olympic-sized swimming pools. No one knows where this water will come from either.

11

ENERGY DEMANDS PLACE AN UNSUSTAINABLE BURDEN ON WATER

Energy and water are tightly entwined. It takes a great deal of energy to supply water, and a great deal of water to supply energy. With water stress spreading and intensifying around the globe, it's critical that policymakers not promote water-intensive energy options. —**Sandra Postel, Global Water Policy Project**[1]

THE RELENTLESS SEARCH FOR new energy and mineral resources is placing an untenable burden on the world's water supplies. As it is used to provide a cheap (and assumed to be limitless) resource for the global food trade, so water is used to promote the exploration and production of energy. As in the food trade, water is both over-extracted and polluted in the hunt for energy resources.

In early 2013 the International Energy Agency (IEA), a Paris-based intergovernmental organization that conducts research and advises member states on energy policy, published a comprehensive report predicting that the volume of water consumed for energy production worldwide would double by 2035. That is four times the volume of the largest U.S. reservoir, the Hoover Dam's Lake Mead.

While conventional energy continues to be a growing problem and the fracking craze is worrisome, the agency says the biggest energy culprits are biofuels and coal-fired power plants.[2]

COAL-POWERED ELECTRICITY DRIVING WATER DEMAND

Coal-power use is increasing everywhere except in the United States, and coal may surpass oil as the main source of energy by 2017, according to the IEA report. Coal is the world's most abundant fossil fuel and is relatively inexpensive. The biggest coal producers are China (by far the biggest, with 50 percent of world production), the United States, India, Australia, Russia, South Africa, and Indonesia. Australia is the largest coal exporter, sending about 70 percent of its production abroad, mostly to Japan. Coal is considered the dirtiest of the fossil fuels, as it releases carbon directly into the air when burned. But coal mining also destroys local watersheds through toxic acid drainage that remains for decades.

To convert coal to electricity, it is milled to a fine powder and blown into a boiler's combustion chamber, where it is burnt at a high temperature. The heat energy created converts water into steam, which is passed into a turbine that rotates at high speed. This turbine rotates a generator of wire coils in a strong magnetic field, creating the electricity. Water is used to extract, wash, and transport the coal; to create and then to cool the steam used to make electricity in the power plant; and to control pollution from the plant. Water for the plant is generally taken from local sources. The Union of Concerned Scientists says that a typical coal plant withdraws up to 700 billion litres a year and consumes up to 4 billion litres of that water.[3]

If today's trends hold, says the International Energy Agency, water consumption for coal electricity will jump 84 percent by 2035, to 70 billion cubic metres annually, and will be responsible for more than half of all water consumed in energy production. Greenpeace says this will lead to a water crisis in South Africa, where the national utility, Eskom, is building two new mega–power stations. The plants are estimated to use 173 percent more water per unit of electricity than wind power, and there are fears that water will be diverted from food production and residential use to meet the demands of the mining industry. South Africa is projected to be experiencing a 17 percent gap between water supply and demand by 2030.[4]

A similar warning has been sounded for China. A March 2013 analysis by Bloomberg New Energy Finance says that roughly 60 percent of China's new coal-fired power plants are located in northern China but only 20 percent of the country's water is found in the north. By 2030 the amount of water used by China's power sector could be as much as 190 billion cubic metres, compared to 102 billion cubic metres in 2010. That would constitute a quarter of the country's water supply in that year. "Given that some regions are already in water deficit today, the projected increase in power-related water withdrawals could quickly become unsustainable," says the report.[5]

BIOFUELS WASTE PRECIOUS WATER

In the past decade the production of biofuels has increased exponentially. Intended to decrease the use of fossil fuels and cut greenhouse gases, the notion of biofuels has found support across the political spectrum. It was a well-intentioned policy move with ill-considered ramifications. In the United States, 40 percent of the corn crop is currently diverted to make fuel for cars. Lester Brown, of the Earth Policy Institute, says that corn would have fed 350 million people.[6] Cornell University professor of ecology and agriculture David Pimentel, who has studied biofuels extensively, reports that corn is the number one cause of soil erosion in the United States, and its overdependence on nitrates, herbicides, and insecticides is the prime reason for the dead zone in the waters of the Gulf of Mexico.[7]

Concerns are also growing about the amount of energy it takes to produce biofuels. The *Guardian's* George Monbiot quotes a UN report on the rapid destruction of the Indonesian rainforest due to clearing for planting palm oil for biofuels over the next decade. As the forests burn, both trees and the peat they grow in are turned into carbon dioxide. He also quotes a Dutch study that shows that every ton of palm oil results in thirty-three tons of carbon dioxide — ten times as much as petroleum produces.[8]

But even more alarming is the amount of water consumed in the production of biofuels. It takes 1,700 litres of water to produce one litre of corn ethanol in the United States. A study published in *Environmental Science and Technology* says that the U.S. congressional mandate to produce 60 billion litres of corn ethanol a year by 2015 would require an estimated 6 trillion litres of additional irrigation water annually, and even more in direct rainfall — a volume that exceeds the yearly water withdrawals of the entire state of Iowa. The highly respected scientist authors note that replacing gasoline with corn ethanol results in significant problem shifting and causes greater damage than gasoline: "Our study indicates that replacing gasoline with corn ethanol may only result in shifting the net environmental impacts primarily toward increased eutrophication and greater water scarcity."[9]

According to the International Water Management Institute, in areas where the only water source for biofuel production is irrigation, the amount of water consumed is even higher. For example, a litre of ethanol in India requires 3,500 litres of irrigation water. In China it requires 2,400 litres.[10] Yet despite its heavy water footprint, the production worldwide of biofuels is growing, having risen in one year by 17 percent, to 105 billion litres, in 2010. The International Energy

Agency expects biofuels to meet more than one-quarter of world demand for transportation by 2020. In its report *Biofuels Markets and Technologies,* Navigant Research, a technology consulting company, predicts steady growth in the biofuels industry through to 2021, when production will reach 260 billion litres.[11]

The International Energy Agency warns that this will place an intolerable burden on the world's available water supplies. In its report on world energy production, the agency anticipates a 242 percent increase in water consumption for biofuel production by 2035, from 12 billion cubic metres to 41 billion cubic metres annually, and says biofuels will account for 72 percent of the water used for primary energy production.[12]

Several recent studies comparing different types of fuels used in cars found biofuels to be by far the worst water guzzlers. Carey King and Michael Webber, of the University of Texas, found that for every kilometre driven, electricity from the grid uses 0.56 litres of water, while gasoline from petroleum uses 1.5 litres of water. But fuel from irrigated corn or irrigated soybeans uses 35 litres of water. And because they are from agricultural products, the biofuels have much larger consumption levels — that is, the water is not returned to the source.[13] Three American scientists published similar findings in a study called "Burning Water: A Comparative Analysis of the Energy Return on Water Invested." The scientists say that developing large-scale biofuels to help counter the world's fossil problem "may produce or exacerbate water shortages around the globe and be limited by the availability of fresh water."[14]

A critique of biofuels in no way endorses the use of fossil fuels or detracts from the urgent need to cut their emissions. It is, however, an urgent wake-up call for the world not to pit air against water and to assume that the latter has the carrying capacity to sustain this level of consumption.

BRAZIL IN THE LEAD

Brazil and the United States produce almost 90 percent of the world's biofuels. Brazil is the leading manufacturer and exporter of sugarcane ethanol, and most cars in Brazil now run on a mixture of ethanol and gasoline. Brazil currently produces 28 billion litres of sugarcane ethanol (2 billion litres of which is exported) and wants to be producing 200 billion litres by 2020, although the U.S. Department of Agriculture predicts that it will likely be closer to 44 billion litres.[15]

It takes a great deal of water to produce this biofuel. Cornell professor Pimentel estimates that it takes 2,655 litres of water if the crop is irrigated (1,720 litres if it is rain-fed) to produce one litre, counting the water used to grow the sugar cane as well as that used in the production process.[16] Scientists A. Y. Hoekstra and P. W. Gerbens-Leenes corroborate this figure in a study for UNESCO.[17] Currently 7 trillion litres of water are extracted every year to produce ethanol in Brazil. In less than a decade this figure could reach an astonishing 65 trillion litres of water.

Biofuel production in Brazil destroys forests, which are cut down to make way for the vast sugarcane fields; threatens the savannahs and the Amazon River Basin; and contaminates the water and soil with chemical fertilizers. Small farmers and indigenous landholders have been forced off their land to make way for the new agribusiness. Two-thirds of the Cerrado, the world's most diverse savannah — it lies between the Amazon and the Atlantic rainforest and is known by the locals as the "Father of Water" — has been degraded for cattle ranching and sugarcane production. The Ipojuca River in northeastern Brazil has been contaminated by nitrate leaching, acidification, and oxygen imbalance from biofuel production. Many rural streams and rivers have dried up as large biofuel farms move in and draw from them.

All this places the Guaraní Aquifer — the world's largest, lying beneath Brazil, Argentina, Paraguay, and Uruguay — at risk. Water is being extracted from the aquifer faster than it can be recharged. High salt levels and reduced water pressure result, possibly rendering the extraction process very difficult in the future. Heavy metals, toxins from the mining and forestry industries, urban sprawl, poorly treated sewage, phosphorus, fertilizers, agro-chemicals, and multipoint contamination combine to spill a witch's brew of poison into the aquifer. Karin Kemper, a senior water resources specialist with the World Bank, says, "The Guarani is a striking example of an international water body threatened by environmental degradation. Without better management, the aquifer is likely to suffer from pollution and rapid depletion. Uncontrolled exploitation could reduce it from a strategic water reserve to a degraded resource that is a focus of conflict in the region."[18]

The big agro-companies are given preferential access to the waters of the region over local needs. Global agribusiness companies such as Cargill, the world's leading sugar producer and trader, and energy giants such as Royal Dutch Shell and BP, are piling into Brazil in anticipation of huge growth in the biofuels industry. Meanwhile, potable water and sanitation services are not reaching at least 25 percent of the people of the region. As elsewhere, it is the indigenous, the poor, the *favela* dwellers who are left behind in the rush to use the region's water resources for export and profit.

The argument made by the Brazilian government, the industry, and biofuel proponents is that Brazil has an abundance of water, so this usage is an acceptable trade-off. But this is a short-sighted view and does not take into account the growing evidence that even regions blessed with water plenitude can go dry if they abuse (and export) their water heritage. Not surprisingly, Brazil has started experiencing something fairly rare: serious, frequent, and prolonged droughts. The country had a fierce drought in 2005 that caused a massive die-off of trees in the rainforests, and another in the fall of

2010 that marked one of the worst on record for the Amazon. The Rio Negro, a major tributary of the Amazon River, dried up. A nineteen-month drought that stretched from 2011 into 2012 reached across Argentina, Paraguay, and Brazil, hitting Brazilian soybean producers hard. In early 2013 the northeast suffered its worst drought in fifty years, threatening hydro power supplies, plunging Rio de Janeiro's airport into darkness, and wiping out 30 percent of the region's sugarcane production. Dams were at one-third capacity, and 20 million people living in the semi-arid region needed government aid.

Scientists point to the destruction of natural vegetation, the pumping of ground- and surface waters, deforestation, and the drying out of the Amazon rainforest as causes of Brazil's droughts, all to some extent caused by the biofuel craze. "Every ecosystem has some point beyond which it can't go," says Oliver Phillips, a professor at the University of Leeds who has spent decades studying tropical forests and climate change. "The concern now is that parts of the Amazon may be approaching that threshold."[19]

DOWN AND DIRTY IN THE TAR SANDS

While coal-fired electricity and biofuels top the list, oil and gas exploration and production will still account for 10 percent of the water consumed for energy production by 2035. And much of this production presents its own unique threat to local water supplies. As the world is running out of conventional oil and gas, it is looking further afield to less conventional and more environmentally dangerous methods, such as deep-sea drilling, oil or tar sands, and fracking. Tar sands are a type of petroleum deposit containing sand, clay, and water saturated with a dense form of petroleum called bitumen. It has the consistency of molasses, and the challenge is to remove the oil from the rest of the mixture. While there are deposits in other countries, over 70 percent of known reserves are found in Canada, most of them in northern Alberta.

The deposits around Fort McMurray, Peace River, and Cold Lake, the largest in the world, lie under 141,000 square kilometres of boreal forest — an area bigger than Scotland — and contain about 1.7 trillion barrels of bitumen, almost 2 million of which are processed every day. The process of extracting the oil from the bitumen is energy-intensive, producing far more greenhouse gas emissions than conventional oil. But the greatest concern is the damage to local water supplies.

It takes huge quantities of water to steam-blast oil from the sands. For every barrel of oil recovered from the tar sands, three to five barrels of water are used, according to the respected Pembina Institute. Currently, approved tar-sands operations are licensed to remove a quantity of water from the Athabasca River that is more than twice the volume required to meet the annual needs of the city of Calgary. It accounts for three-quarters of the river's water, and its summer flows have declined by 30 percent since 1970.

After it is used, the water is dumped into massive toxic lagoons so dangerous that birds die on impact. One hundred and seventy square kilometres of these poison lakes leach 11 million litres of toxic water into the watershed every day, say David Schindler and his colleagues. The government of Alberta and the big energy companies have come up with a $40-billion plan of expansion for the tar sands. If they are successful, the operation will be producing about 5 million barrels per day of the dirtiest oil on earth, which could in turn use — and destroy — about 20 million barrels of water per day.

This expansion would require an additional 14,000 kilometres of pipeline, raising deep concern over potential spills. There have been more than four thousand oil spills in Canada between 2007 and 2009 alone.[20] On July 25, 2010, an Enbridge pipeline ruptured in Michigan, spilling almost 4 million litres of Alberta tar-sands bitumen (diluted with chemicals to help it move through the pipes) into the Kalamazoo River. Damage to the river, its estuaries, and its aquatic life was widespread, and a number of families had to be evacuated. By summer 2012 the clean-up expense to the Michigan and U.S. governments had come to $765 million. Enbridge was fined $3.7 million by the U.S. Department of Transportation.[21]

These spills, the threat to the environment, and the destruction of Alberta's boreal forest have brought together a powerful alliance opposed to the construction of any new pipelines. First Nations are leading the opposition in many communities. Aboriginal people living near the tar sands report unusually high rates of rare cancers and autoimmune diseases; they cannot hunt local game or drink the water from traditional sources. They report fish with three eyes and two mouths. I have worked with the Athabasca Cree Nation of Fort Chipewyan and others in their legal and community fights against oil development, and I can attest to their commitment to protect the environment by having these operations phased out. I toured the region with native activists and spoke alongside youth who said it would be better for the government to come in and shoot them with guns rather than take them down one by one with cancer.

In British Columbia — dubbed by some "Canada's carbon corridor" for all the pipelines, gas terminals, coal mines, and fracking operations either in place or planned — sixty-two First Nations, the Union of B.C. Municipalities, and a huge coalition of farmers, environmentalists, homeowners, health-care providers, and others have come together to say no to the transporting and export of Alberta tar sands oil through their province. Their major goal is to stop construction of the Northern Gateway Pipeline, which would carry the bitumen to markets in Asia across eight hundred pristine rivers and streams, loaded onto supertankers that carry more oil each than the *Exxon Valdez*.

Opposition to the Keystone XL Pipeline, which would carry Alberta bitumen to refineries in Texas, arose in the United States because of the threat of spills to the Ogallala Aquifer, over which the original route would have passed. I attended many of the "arrest" sessions in front of the White House led by 350.org's indomitable Bill McKibben and stood side by side with American youth as we surrounded the White House four times over. In September 2011 I was arrested, with several hundred others, in front of the Parliament Buildings in Ottawa after crossing a police barrier in a protest against Keystone; I was escorted off Parliament Hill in a paddy wagon — one of the proudest days of my life. If the tar sands are allowed to proceed as planned, northern Alberta will eventually be the site of the largest greenhouse gas emissions in the world. Doing my small part to prevent this meant that I could look my grandchildren in the eye the next time I saw them.

The government of Stephen Harper is determined to turn Canada into an energy superpower and is removing all impediments. Prime Minister Harper abandoned the Kyoto Protocol, making Canada the only country in the world to ratify and then abandon that climate treaty. He defanged energy conservation programs and all the major environmental laws that protected Canada's freshwater heritage, including the Fisheries Act and the Navigable Waters Protection Act. He gutted the Environmental Assessment Act, cancelling more than three thousand environmental assessments already underway. He cut funding for many important water-protection programs inside government, and key research projects and facilities that conducted independent research; these included the Experimental Lakes Area, a site for freshwater experiments on acid rain and eutrophication, as well as the Global Environment Monitoring System Water Programme that Canada ran for years. This research network monitored the health of freshwater lakes around the world for the United Nations. And in March 2013 Stephen Harper pulled Canada out of the United Nations Convention to Combat Desertification, which Foreign Affairs Minister John Baird called a "talkfest."

This assault on Canada's freshwater heritage and on Canada's role in finding solutions to the world's water problems is deeply disturbing. At a time when all governments and institutions of influence should be coming together to face and confront the water crisis, too many are turning a blind eye.

FRACKING ON THE RISE

The International Energy Agency had noted the surge in fracking around the world and says that, by 2035, water consumption for natural gas production will increase by 86 percent.[22] While that is a large increase, the overall water footprint of fracking does not come near that of biofuels and coal-fired hydroelectricity. But the impact of fracking operations on local watersheds can be severe. For instance, in the shale-gas operations in Pennsylvania, 15 million litres of water are required for each fracked well, compared to the 378,540 litres for a conventional gas well.[23]

Fracking, short for "hydraulic fracturing," is a water-intensive process in which a mixture of sand, chemicals, and water is injected deep underground at high pressure to release natural gas from rock formations. It uses large amounts of water from local sources and contaminates that water with the chemical cocktail used in the process. While the chemicals used are under trademark protection and are therefore considered "trade secrets," a 2011 American study identified more than six hundred chemicals used in fracking, at least a quarter of which have been linked to cancer and mutations and half of which can affect the nervous, immune, and cardiovascular systems.[24] Fracking also releases methane gas, which contaminates local wells and water supplies and adds to greenhouse gas emissions when released into the air.

There are reports of livestock illnesses and deaths on farms near fracking operations, as well as reports of earthquakes near sites of intense shale-gas fracking. Canadian journalist Andrew Nikiforuk reports that both EnCana and Chesapeake Energy, two of the largest shale-gas players, have assembled land bases equal in size to the state of West Virginia for shale-gas drilling alone.[25]

The U.S. government estimates that the biggest reserves are in China, the United States, Argentina, Mexico, and South Africa. None have the water reserves to support this industry, but that has not stopped them. A report by the Beijing independent news service Caixin says that China is ratcheting up its fracking ambitions with no regard for groundwater protection. The government admits that it will likely take up to five years to write rules to deal with the environmental effects of fracking, but it has given the green light to industry to implement full-scale exploration nonetheless. Even when the rules are ready, they will likely not be binding.

The Chinese government notes that fracking in the United States grew fourteen times in its first decade. It hopes to follow that lead and to be withdrawing 6.5 billion cubic metres of gas a year by 2015 and 100 billion cubic metres by 2020. To meet the 2015 goal will require an estimated 13.8 million cubic metres of water, just under half of the water now used by China's entire industrial sector.[26]

The fracking industry is aware of the limitations that water scarcity places on its future. In an advertisement for a conference promoting fracking in Argentina, the corporate sponsors — which include energy and fracking companies, the U.S. embassy in Argentina, and the International Private Water Association — note that Argentina has vast reserves of shale gas. But they recognize that they have a problem, because the water supplies it would take to access these reserves are daunting: 140 billion litres, an amount of water equal to the total quantity supplied daily by public systems to the whole country. Not surprisingly, "shale water management" features largely in this conference.

South Africa is also in the midst of a planned fracking boom after the government lifted a 2011 ban on exploration in 2012. Concern about water supplies has grown in the Karoo, an ecologically sensitive semi-arid region of the Eastern Cape, under which most of the natural gas lies. Royal Dutch Shell, Falcon Oil and Gas, and Sunset Energy have been granted licences to explore for gas.[27]

Over the past decade, fracking has exploded in the United States, especially in the Midwest, where the flares and lights from operations can now be seen from space. Supported by the Obama administration as a way to wean the country off foreign oil supplies, fracking has brought environmental devastation and declining property values to rural communities. As Food and Water Watch reports, the industry is exempt from key federal water protections, and federal and state regulators have allowed unchecked expansion of fracking across many states. This has polluted drinking-water supplies, contaminated groundwater, and spilled into rivers and lakes. A 2011 Food and Water Watch report cites a variety of studies confirming that fracking companies have injected millions of litres of chemicals such as diesel fuel — which contains the known carcinogen benzene — into water systems. The amount of benzene from a single fracked well could contaminate almost 400 billion litres of drinking water.[28]

The biggest fracking operation in the world is generally agreed to be the EnCana operation in the Horn River Basin of northwestern British Columbia, with sixteen five-acre well sites and twenty-acre well pads. It has turned an entire wilderness area into an industrial site in just a few years. Little regulation or public consultation has led to rampant development in northern B.C. The provincial government offers royalty credits for every well drilled. The Fort Nelson First Nation is deeply concerned about plans to expand fracking in the area that would dam the Peace River, flooding hundreds of square kilometres. The provincial government also has plans to approve dozens of new licences that would allow energy companies to take billions of litres of water from local lakes and rivers. The band calls the project the "Shale Gale" and has stated that it represents "the largest and most destructive industrial force that our waters have ever known."[29]

The Canadian Centre for Policy Alternatives reports that the B.C. government has plans to double or even triple the amount of natural gas produced in the province, much of it fracked, and will contribute greenhouse gas emissions equivalent to putting at least 24 million new cars on the road. The government is essentially giving the industry twenty-year access to public water supplies, with little or no public input despite the fact that the project uses up to six hundred Olympic swimming pools' worth of water per gas-well pad.[30]

NUCLEAR ENERGY

While the International Energy Agency's study says that the nuclear industry's water footprint is small in comparison to those of coal-fired electricity, biofuels, and oil and gas, this does not mean that nuclear power does not threaten water. Nuclear power requires a lot of water for the cooling process, which is why nuclear plants are built beside lakes or oceans, and the water it returns to the watershed is approximately 25 degrees warmer than it was originally. Damage is also done to groundwater when radioactive nuclear waste is stored and the waste leaks.

The greatest threat of nuclear power, however, is the possibility of an accident in which some of the most dangerous elements known to humans spill into local water systems, contaminating them for decades or longer. The Chernobyl disaster contaminated vast bodies of water in Ukraine, Belarus, and western Russia, and affected waterways as far away as Poland, Norway, and Sweden.

RENEWABLE ENERGY

The destruction of the planet's water for our energy demands is a largely untold story. The increase in greenhouse gas emissions has been widely studied and documented, and most people now understand that these emissions cause serious climate instability. However, the public and most opinion leaders and elected officials still have no idea that the exponential growth in all the conventional forms of energy, as well as biofuels, is placing the planet's dwindling supplies of accessible water in grave danger. The need to find and support sustainable alternatives to current energy sources is urgent if we are to save the world's water.

This goal is entirely possible, says the World Future Council, a Hamburg-based organization of fifty international "councillors" founded by Jakob von Uexkull, a philanthropist and former member of the European Parliament, to bring the interests of future generations to the centre of policymaking. Stefan Schurig, who heads the council's climate and energy department, says that more than 50 percent of all energy investments in the world are now in renewables and that it is entirely possible to achieve what he calls the "100 percent target" using conservation and wind, tidal, biomass, and solar power. Already, notes Schurig, 132 regions in Germany have committed to the 100 percent target, mostly through feed-in tariffs through which home and business owners are paid for generating their own electricity using solar power. Over 20 percent of Germany's electricity now comes from renewables, saving 87 million tons of carbon dioxide a year.

However, it is not enough just to look at the positive impact of renewable energy on the air. We must also look at its impact on water, and there are some very real concerns about solar power and the water it takes to produce it. The Union of Concerned Scientists reports that while individual solar cells and panels do not use any water for generating electricity, the larger utility-scale solar thermal plants called concentrated solar power (CSP) systems need large amounts of water for cooling. They need the water because most of them, while using the sun as fuel, generate power to produce electricity by creating steam to turn turbines. Once the steam has done its job, it has to be cooled down so the cycle can start again. Some large operations require as much or more water than the coal, natural-gas, or nuclear power plants they are meant to replace. Two power plants being built in the Mojave Desert would require almost 6 billion litres of groundwater annually.[31]

These utilities can cut their water use by up to 90 percent by using dry cooling, which blows air to cool the steam generated from the process, rather than wet cooling, which uses water to cool the steam back into water, on top of the use of grey water to meet steam generation needs. Such practices cost more and are therefore often not adopted. However, policymakers need to take water into account when they set up rules for solar and other new energy sources. The point to be made here is that, in our search for alternatives to fossil fuels, we do not sacrifice water for air. Both must be protected.

12

PUTTING WATER AT THE CENTRE OF OUR LIVES

Think like water.

— Denise Hart, Save Our Groundwater, New Hampshire

S AVE OUR GROUNDWATER (SOG) is a grassroots volunteer organization formed in 2001 by residents of the southeastern New Hampshire communities of Nottingham and Barrington. Their goal was to fight an application by a bottled-water company to withdraw 1.6 million litres a day from the local aquifer, and the fight against USA Springs took eleven years. It required intensive research and documentation, political lobbying, community organizing, and a legal challenge that went all the way to the state Supreme Court. Although the group encountered one setback after another, they persevered. The company filed for bankruptcy in 2008 and the federal bankruptcy court took control of its dissolution in 2012. The groundwater of southeastern New Hampshire is safe for now.

Denise Hart is a soft-spoken environmental and human rights advocate and one of the leaders of this coalition. I have worked closely with Denise, her husband, Michael, and the SOG team, and

visited the area many times. I once asked Denise what she meant by
the words "Think like water."

She responded, "When I say this, I think of how water confronts
obstacles to its path — it never gives up, but instead goes around,
pushes over, changes course, all the while continuing on its way.
Water has taught me how all life is interconnected, how we all relate
to each other. Rivers are connected to lakes and springs; lakes are
often fed by groundwater and so on. What happens upstream affects
all downstream. It is a different way to think about the world. Water
celebrates our connectedness and interdependence. If we humans use
up the water or foul it, our communities and the wildlife, plants, sea
life, all are affected. So, we have to learn to think more like water and
come to know the systems of water, which sustain life in abundance
if we use it wisely and sustainably, thinking ahead for the well-being
of our neighbours and future generations of all life."

A NEW WATER ETHIC

We humans have allowed the planet's fresh water to be used as a
resource for the modern world we have built, rather than seeing it as
the essential element in a living ecosystem. It seems very clear to me
that we need to change our relationship to water, and we need to do it
quickly. We need to find out what makes water sick and what makes
it well again and to do all in our power to heal and restore the waterways and watersheds of our ecosystems. Not only must we reject the
market model for our water future, we must also put ourselves at the
service of undoing what we have done to the natural world and hope
that it is not too late.

It is time to learn some humility. We must adopt a new water
ethic that puts water protection and restoration at the centre of the
laws and policies we enact. What would our cities look like if we no

longer paved over rivers and streams but instead built around and celebrated them? What would agriculture policy look like if we had laws with teeth (as they do in northern Germany), preventing food-producing activities from harming the local water systems? What would trade policy look like if the true costs of virtual water loss were factored into the cost of production? What would energy policy look like if we considered the destruction of fresh water? How would we look at water diversion and dams if we accepted that rivers need to flow to remain healthy?

Conservation is a key component of a water ethic and relatively easy for us to adopt. When my grandchildren turn off the tap as they brush their teeth, I know they are being taught to care for water. The Global Water Policy Project's Sandra Postel, who has been sounding the alarm about water for decades, says that measures to conserve, recycle, and more efficiently use water have enabled many places to contain their water demands and avoid — even if only temporarily — an ecological reckoning. She notes tried-and-true measures such as thrifty irrigation techniques, water-saving plumbing fixtures, investment in infrastructure to stop water loss through leaking pipes, native landscaping, and wastewater recycling as cost-effective ways to reduce the amount of water required to grow food, produce material goods, and meet household needs. She adds that the conservation potential of these measures has barely been tapped.

But something is still missing from this prescription, she argues in an essay for the *American Prospect,* something less tangible than low-flow showerheads and drip irrigation. That something has to do with modern society's disconnect from nature and from water's fundamental role as the basis of life. "In our technologically sophisticated world, we no longer grasp the need for the wild river, the blackwater swamp, or even the diversity of species collectively performing nature's work....Overall, we have been quick to assume rights to use water but slow to recognize obligations to preserve and protect it." She says the essence of a water ethic is to make the protection of freshwater ecosystems a central goal in all that we do.

The adoption of such an ethic would shift human activity away from the strictly utilitarian approach to water management towards an integrated, holistic approach that views people and water as interconnected parts of a greater whole. "Instead of asking how we can further control and manipulate rivers, lakes, and streams to meet our ever-growing demands, we would ask how we can best satisfy human needs while accommodating the ecological requirements of freshwater ecosystems," Postel argues. This would lead us to deeper questions of human values, "in particular how to narrow the wide gap between the haves and have-nots within a healthy ecosystem."[1]

Canadian geologist and writer Jamie Linton promotes the concept of the "hydrosocial cycle," a process in which flows of water reflect human affairs and human affairs are enlivened by water. "The task, already begun, is to put the hydrosocial cycle to work in helping promote social equity and environmental sustainability not just in cities, but wherever intervention in the hydrologic cycle has produced inequitable or uneven access to water and water services."[2]

WATERSHED RESTORATION IS CRUCIAL TO RECOVERY

Essential to recovery of climate stability and a water-secure future for all is the conservation, protection, and restoration of watersheds. We must stop mining water from lakes, rivers, and aquifers and rebuild the health of our aquatic ecosystems if the planet and we are to survive. Slovakian hydrologist and Goldman Prize winner Michal Kravčík is leading a global effort to save the earth's hydrologic cycle with watershed restoration. His groundbreaking research has shown that when water cannot return to fields, meadows, wetlands, and streams because of urban sprawl, poor farming practices, overgrazing, and removal of water-retentive landscapes, the actual amount of water in the local hydrologic cycle decreases, leading to desertification of once green land.

When we remove water from the soil, it heats up and heats the air around it. Similarly, when we remove vegetation (wetlands, forests, native grasses) from the soil, water vapour is lost to the local water-shed, which in turn causes climate warming. A team of scientists led by University of Leeds researcher Dominick Spracklen have found that air passing over large rainforests produces twice as much rain (or more) as air that passes over deforested areas. Trees release moisture in the process of evapotranspiration, the scientists explain, which keeps the air moister and cooler, lowering temperatures. The moist air can travel for days, bringing needed rains thousands of kilome-tres away. The team estimates that, at current trends of deforestation in the Amazon alone, rainfall in the Amazon Basin will decline by 12 percent in the wet season and 21 percent in the dry by 2050.[3]

Kravčík says that we are losing vast amounts of water from the hydrologic cycle to the oceans every year through this abuse of watersheds and forests, and he believes that our abuse of water is the leading cause of climate change. He convinced the Slovakian government to implement a national watershed restoration plan, based on traditional methods of rainwater harvesting through stor-age reservoirs; small wooden, stone, and earthen "holding" dams; cross drains and infiltration pits on country roads; and reparation of eroded ravines and gullies. The first phase of the project, which ended in 2011 and involved 190 municipalities and more than fifty thousand water-retention measures, was a huge success and helped recover large areas of degraded land. The project, called Blue Alternative, also employed almost eight thousand people, many of whom — including the large number of Roma people involved — had previously been on social assistance. Slovakia has eagerly embarked on the second phase of this project.[4]

Rainwater harvesting is the collection and storage of rainwater, and it has been used in arid and semi-arid areas for millennia. The technique is now being employed in urban areas as well. One method suitable for cities is rooftop harvesting, which can be a source for recharging groundwater. China and Brazil have extensive rooftop rainwater-harvesting programs, and Bermuda has a law that requires all new construction to include rainwater-harvesting facilities.

A similar experiment to Slovakia's has been created in the peace research village of Tamera, in southwestern Portugal, where a community of visionaries has used rainwater harvesting and permaculture farming techniques to transform a desertified landscape — created by bad forestry and farming practices — into a green paradise. Ten lakes surrounded by healthy wetlands and marshes have been created to retain the water of the rainy season and give it time to soak into the ground. The community is farming successfully, and biodiversity, wildlife, and young forests are all returning to the region.

The Centre for Science and Environment in New Delhi, India, runs dozens of rainwater-harvesting programs around the city, and has trained thousands from all over the country to renew this ancient technique for water retention. In Rajasthan, Rajendra Singh's Tarun Bharat Sangh movement has brought life and livelihoods back to the region through a system of rainwater harvesting that has made deserts bloom and rivers run again. Singh says that if the drops come from the clouds, humans can catch them. Singh's organization and the villagers of the area have built more than ten thousand water-harvesting structures in the past twenty-five years and have revived seven rivers across the state. People come from all over the world to learn from Singh, who is known in India as "the rain man"; his work and vision have brought health and harmony to hundreds of once rain-impoverished communities.

Adelaide, the capital of South Australia, lives with the uncertainty of erratic water flows from the Murray and Darling Rivers. In some seasons the rivers do not reach the ocean; in some seasons the riverbanks overflow. Unfortunately, even when rainfall is plentiful, most of the floodwater is not captured and runs into the sea. While the city has opted for a "modern water" solution — desalination and privatization of water services — a resourceful and determined former farmer named Colin Pitman has implemented a natural solution to water scarcity and water pollution in the Adelaide suburb of Salisbury.

Pitman convinced the Salisbury Council to let him create fifty-three wetlands on the Salisbury Plains, where he purifies storm water by using carefully selected species of water plants, freeing it of 90 percent of the potassium, phosphorus, and heavy metals it carries. He then stores the water in aquifers for later retrieval for industrial use and irrigation. Salisbury has gained international recognition for the way it harvests urban runoff and purifies polluted water; government officials and scientists come from all over the world to learn how to green their deserts as Salisbury has done.

Pitman says that Adelaide could supply its citizens with much cheaper and more reliable water than from its expensive and controversial desalination plant. The local newspaper expressed outrage at Adelaide's refusal to learn from Pitman:

> When it rains in Salisbury, the council makes money, businesses reap water, and the environment is protected.
>
> When it rains in the rest of Adelaide, a huge amount of polluted water sweeps out to sea where it poisons and stifles sea grasses, strips beaches of sand, damages the environment, and does nothing for our chronic water shortage.
>
> Adelaide is facing a crisis with its water supply, thanks to drought, an over-reliance on the River Murray and, according to critics, a lack of foresight. In stark contrast, Salisbury, one of

Adelaide's city councils, has pioneered a bold new approach that is being monitored around the world.[5]

And then there is the beauty of the network of wetlands where there was once only desert. When I toured Pitman's project, I was stunned at the bounty of aquatic life. One hundred and seventy bird species have been recorded in these wetlands, many of them not seen for years.

The United Nations gave its 2013 "Water for Life" Best Practices Award to the city of Kumamoto in Japan, located on Kyushu, the southern major island. The city is blessed with abundant groundwater in the volcanic aquifer created by Mount Aso, and the residents are determined to pass on this quality of water to future generations. As with Colin Pitman's project, the city manages a groundwater-recharging system that uses abandoned paddies and protected forests. The water is so pure the people call it "mineral water from the tap."

(These groundwater recharge projects are not to be confused with groundwater dumping. As of July 1, 2013, Florida's new law permits more toxic waste and sewage to be dumped into the state's groundwater. Local municipalities that banned toxic waste now have to accept asphalt, combustible petroleum waste, cement products, plastic paints, insulation and other poisonous chemicals. Mining, construction, and oil and gas companies no longer have to incur costs to safely dispose of their wastes, and can freely pour them into ancient aquifers.)

Some local governments are waking up to the need to protect and restore watersheds, realizing that it is far more cost-effective to invest in ecosystem restoration than in the kind of "modern water" engineering they have favoured in the past. In 2011, according to a report of the American NGO Forest Trends, local governments around the world invested more than $8 billion in watershed protection projects, recognizing that trees, wetlands, and grasses are extremely effective at cleaning and retaining water, as well as reducing sedimentation that clogs water reservoirs.

In China, reports Stephen Leahy of Inter Press Service, residents of struggling communities upstream of the southern coastal city of Zhuhai receive health insurance benefits as an incentive for adopting land and soil management practices that protect drinking water. In Bolivia's Santa Cruz Valley, more than five hundred families receive beehives, fruit plants, and wire for fencing to keep livestock away from rivers and stream banks. A Swedish water authority supports a program to establish blue mussel beds in a fjord to filter nitrate pollution, rather than using a chemical-based treatment plant.

Instead of using electric-powered water treatment plants, New York City brings in its high-quality drinking water through aqueducts connected to protected areas in the nearby Catskill/Delaware forests and wetlands. The city has saved between $4 billion and $6 billion on the cost of water treatment by protecting forests and compensating farmers in the Catskills for reducing pollution in their lakes and streams. New York City now uses the sun's power to ensure clean drinking water, through a new state-of-the-art water disinfection facility that uses ultraviolet radiation to destroy waterborne pathogens. Water officials hope to phase out or at least cut the use of chlorine.[6]

A number of countries and communities are realizing that they need to recapture old knowledge about how nature can protect them from unexpected storms and rising waters. In Thailand there is a concerted effort to rebuild the mangrove forests that once lined coastal areas and offered protection from tsunamis. And in the wake of the damage done by Hurricane Sandy, New York governor Andrew Cuomo has proposed a plan that would turn properties in Queens, Brooklyn, and Staten Island into parks, bird sanctuaries, and dunes that could act as buffer zones.

New York Times reporter Michael Kimmelman says that Cuomo could learn from the Netherlands. After decades of trying to control rising seas, the Dutch are "starting to let the water in," contriving to live with nature rather than fight a losing battle. Homes are evacuated to create floodplains; offices are built on dykes; public squares are designed to double as catch basins for rain and floodwater; lakes long ago dried up for farmland are recuperated; marshes and wetlands are restored.[7] Their famous Room for the River program is intended to provide flood control by allowing Dutch rivers to expand naturally in times of higher flows. This in turn is greatly improving the environment of water catchments and aquatic life along riverbanks.

Yet so much remains to be done. These examples serve as signposts for a way forward and encourage us to realize that it can be done. The United Nations says that rainwater harvesting in Africa could provide enough water for human consumption, with enough left over to help recharge its 677 major lakes, every one of which is experiencing unprecedented deterioration. Overall, the quantity of rain falling across the African continent is equivalent to the needs of 9 billion people — more than the current global population. Kenya, often considered a water have-not country, has enough rainfall to supply the needs of six to seven times its population. Africa is not water-scarce, says the United Nations Environment Programme; the problem is that the rains come in cycles and most does not get captured and stored.

In the past, rainwater capture and harvesting were common across Africa, but these practices have been replaced by a modern water mentality of draining and diverting lakes, rivers, and aquifers. The UNEP notes that what is needed is not large-scale solutions but low-cost harvesting technologies that are simple to deploy and maintain. "Conserving and rehabilitating lakes, wetlands and other freshwater ecosystems will be vital," says UNEP executive director Achim Steiner, who adds that large-scale infrastructure can often bypass the needs of poor and dispersed populations. Widely deployed, he says, rainwater harvesting can act as a buffer against drought while replenishing damaged watersheds and sustaining healthy river flows.[8]

RECOGNIZING THE RIGHT TO A HEALTHY ENVIRONMENT

These and other water restoration projects show a growing understanding that human well-being depends on a healthy environment and requires a new environmental and water ethic. We are beginning to bring this ethic into law and practice. Leading Canadian environmental lawyer and academic David R. Boyd promotes codifying the right to a healthy environment in law and has documented progress made to date around the world. Since the dawn of the modern environmental era in the 1960s, says Boyd, recognition of the essential connection between human rights and a healthy environment has steadily increased. As of 2012, at least 92 percent of the world's countries recognize the right to a healthy environment through their constitutions, court decisions, laws, or international treaties and declarations.

Three-quarters of world's constitutions now include explicit references to environmental rights and/or environmental responsibilities, says Boyd. He notes that no other social or economic right has achieved such a broad level of recognition in such a short time, and that overall this development reflects a rapid and worldwide evolution of human values. There are a number of examples. After a heated debate, France banned fracking, based on the right to a healthy environment. Sweden has set a goal for sustainability at home using practices that do not harm human health or the environment in other countries; it has become the first country in the world to measure its virtual water footprint to determine its impact on the countries it trades with. This must become standard practice everywhere. No country can claim water sustainability unless it examines its footprint in the parts of the world that grow its food and produce industrialized goods that use a lot of local water.

Boyd, who teaches at Simon Fraser University in British Columbia, loves the story of Beatriz Mendoza, an Argentine healthcare worker who took on all three levels of government and forty-four of the country's most powerful corporations in a lawsuit. Its intent was to force them to begin cleanup of the fouled Matanza-Riachuelo river basin, one of the most polluted places on earth. Based on her constitutional right to a healthy environment, Mendoza charged the governments with ecological mismanagement, and she won her case in the Supreme Court of Argentina in 2008. All three levels of government are obliged to spend a combined \$1 billion in cleanup costs, including new drinking-water treatment plants and social housing for former residents of riverside slums.

Boyd conducted a comparative statistical analysis, which demonstrated that nations with constitutional environmental rights and responsibilities have smaller ecological footprints, rank higher on comprehensive measurements of environmental performance, and have made faster progress in reducing air pollution and greenhouse gas emissions. Interestingly, it is a cluster of English-speaking market economies — Canada, the United States, the United Kingdom, and Australia — that are the holdouts in the trend to recognize the right to a healthy environment.[9] Both the U.S. and Canada participated in the UN's Aarhus Convention, which promotes public participation and access rights to government decisions on environmental matters, but they walked away when the Europeans insisted on including the right to a healthy environment.

Exciting as this new development is, most constitutional amendments are still situated within a human-centric framework. That is, they refer to the right of citizens to live in a clean and safe environment and clarify that it is the duty of governments to provide this environment. But some are taking it further and beginning to create a framework to protect nature itself. In 2008 Ecuador became the first nation in the world to recognize the rights of nature, and a year later Bolivia followed suit. The Ecuadorian constitutional amendment says that nature is entitled to "full respect, existence, and the maintenance and regeneration of its vital cycles, structure, functions, and evolutionary processes." It also says that nature is entitled to restoration and that the state will apply the precautionary principle to activities that could lead to extinction, destruction of ecosystems, or the permanent alteration of natural cycles.

Esteemed Uruguayan writer and historian Eduardo Galeano promotes Ecuador's rights-of-nature law. He says that since the days when "the sword and the cross" made their way into the Americas, Ecuador has suffered repeated devastation, including massive pollution of its Amazon forests by American oil companies. The new amendment is a step towards recovering the ancient native tradition of reverence for nature, which was seen by the Europeans as the sin of idolatry and punished by torture and death. In an anthology on the rights of nature, Galeano writes, "Nature has a lot to say, and it has long been time for us, her children, to stop playing deaf. Maybe even God will hear the cry rising from this Andean country and add an eleventh amendment, which he left out when he handed down instructions from Mount Sinai: 'Love Nature, which you are a part of.'"[10]

Ironically, local indigenous, human rights, and environmental groups used this very law to challenge a contract the government of Ecuador had signed with a Chinese company in March 2012, for a new open-pit copper, gold, and silver mining project in the Condor Highland, in southeastern Ecuador. The groups launched a lawsuit that claimed the mine would violate the protected rights of the water and ecosystems guaranteed in the Ecuadorian constitution. Unfortunately the lawsuit was rejected by the Ecuadoran trial court in early 2013.

But this concept is also being codified in the courts. In a landmark 2012 case, the Whanganui River in New Zealand became a legal entity with its own legal voice under an agreement signed between the government and the iwi (Maori social unit) of Whanganui, an indigenous community with strong cultural ties to the river. The deal followed a protracted court battle over the river. Sandra Postel, writing in *National Geographic,* says that this deal will give further impetus to the idea that nature has rights that should be legally protected, just as people do.[11]

HONOURING PACHAMAMA

In April 2010, following the spectacular failure of the December 2009 Copenhagen climate summit, Bolivia's president Evo Morales brought climate-justice activists from around the world to gather in Cochabamba and create an alternative vision for the future. President Morales expected several thousand guests; instead, more than 32,000 of us descended upon his haunting, landlocked country. Bolivia has a large indigenous population and has held on fiercely to its culture, dress, music, and spiritual connection to nature, or Pachamama ("Mother Earth"). So it is not surprising that out of this gathering came a call to protect nature differently, by recognizing its inherent rights.

The Universal Declaration of the Rights of Mother Earth was proclaimed on April 22, 2010. A number of us presented it in person to UN Secretary-General Ban Ki-moon, who seemed genuinely taken with the notion. The declaration recognizes that earth is an indivisible living community of interrelated and interdependent beings with inherent rights. It defines fundamental human responsibilities in relation to other beings and to the community as a whole. Although many of us worked on it, its principal author was Cormac Cullinan, a South African environmental and human rights lawyer and a leading voice in the movement to have nature protected in law.

Cullinan believes that future generations will look back on ours and view our relationship with nature as a form of slavery. He says the day will come when "the failure of our laws to recognize the right of a river to flow, to prohibit acts that destabilize Earth's climate, or to impose a duty to respect the intrinsic value and right to exist of all life" will be as reprehensible as allowing people to be bought and sold. He grew up and studied law in apartheid South Africa and saw first-hand how the state can use law as a form of social control. He asserts that our legal systems for regulating human behaviour are not protecting the earth because they were not meant to. In fact, he says, our legal and political establishments perpetuate, protect, and legitimize continued degradation of the earth by design, not by accident.

In his landmark book *Wild Law*, Cullinan explains that most legal systems view nature as property, and most laws to protect the environment and other species regulate only the amount of damage that can be inflicted by human activity. He calls for laws that regulate humans in a manner that allows other species to fulfil their evolutionary role on the planet. Human laws and governance systems must promote human behaviour that contributes to the health and integrity not only of human society but also of the "wider ecological community." This does not mean that one could not catch a fish, but that fishing a species to extinction would violate the law. Nor does it mean there could be no commercial activity on a river. However, if the river is dammed and over-extracted to the point where it no longer flows, it is no longer a river; its inherent rights have been removed.[12]

Shannon Biggs, director of the Community Rights program at

Global Exchange in San Francisco, works with communities con-
fronted by corporate harms to assert their right to protect local
watersheds and to enact binding law that places the rights of communities and nature above the claimed rights of corporations.
More than 140 communities across the United States have used
this new understanding to assert their rights to make governing
decisions where they live. In 2006, residents of Tamaqua Borough,
Pennsylvania, adopted an ordinance recognizing natural ecosystems
as "legal persons" for the purpose of stopping the dumping of sewage
sludge on wild land. Citizens in communities across New England
have asserted their right to protect their water sources through a
series of ordinances to prevent bottled-water companies from setting
up operations. In 2010, residents of Mount Shasta, California, successfully campaigned to have an ordinance placed on the ballot that
prevented cloud seeding and bulk water extraction within city limits.

In April 2013, Santa Monica, California, adopted a "Sustainability
Rights Ordinance" that lays out the rights of all residents to clean
water from sustainable sources, clean air from renewable energy
sources, and a sustainable food system that provides healthy, locally
grown food. It also recognizes that the environment possesses fundamental, inalienable rights to exist and flourish in Santa Monica.
The ordinance was the brainchild of Linda Sheehan, executive director of the Earth Law Center in California, which advocates for laws
and policies that recognize and promote the inherent rights of nature
to exist, thrive, and evolve.

Shannon Biggs explains that the law in most Western countries
recognizes the rights of corporations while denying communities the right to protect their own health, safety, and welfare in the
places where they live. Entire human societies have come to value
"endless more," to the detriment of all. Biggs sees the move to assert
local democratic control as a new civil rights movement for people
and the planet. "Passing local laws that assert communities' right to

make governing decisions that affect them directly, and recognizing nature's own right to exist, flourish and regenerate its vital cycles is essentially civil disobedience through local law-making," she says. Our task is to restructure the global economy into many local economies, based on the needs of the biosphere. When this happens, "communities will become true stewards of their ecosystems, protecting and upholding these natural rights."[13]

IS IT TOO LATE?

Can we do this? Can we stop the devastation of our lakes, rivers, streams, and groundwater? Can we think like water? Up to 90 percent of waste water in the Global South flows untreated into waterways and coastal zones. Globally, every day, almost 2 billion kilograms (2 million tons) of sewage and industrial and agricultural waste are discharged into the world's water, equivalent to the weight of the entire human population of 7 billion people. The amount of waste water produced annually equals about six times more water than exists in all the rivers of the world.[14]

The great Victorian English poet Gerard Manley Hopkins said,

And for all this, nature is never spent;
There lives the dearest freshness deep down things.[15]

Let's hope he was right.

WATER CAN TEACH US HOW TO LIVE TOGETHER

This principle recognizes the potential for conflict if the current political and economic framework of competition, unlimited growth, and plunder of land and water for profit is not challenged. Water disputes are looming around the world: between nations, between rich and poor, between small farmers and agribusiness, and between thirsty megacities and rural and indigenous communities. But just as water can be the source of disputes, conflict, and even violence, water can bring people, communities, and nations together in the shared search for solutions. Past histories and differences must be put aside when survival of a threatened shared watershed is at stake. Water survival will necessitate more collaborative and sustainable ways of producing energy, growing food, trading across borders, and producing goods and services. And it will require more robust democratic governance as well as more local control over local water sources. Water will be nature's gift to humanity to teach us how to live more lightly on the earth and in peace and respect with one another.

13

CONFRONTING THE TYRANNY OF THE ONE PERCENT

In elevating the economy above all else, we fail to ask the most important questions: What is an economy for? How much is enough? Are there no limits? We're not asking the critical questions. —**David Suzuki, Canadian environmental leader**

BY MID-2013 THERE WERE 1,426 billionaires in the world, up from 111 in 2000. Mexico's telecom magnate Carlos Slim tops the list with an estimated worth of $73 billion. Next in line is American Bill Gates, with a fortune of $67 billion. Spain's Amancio Ortega, worth $57 billion, is third in line, having made an additional $19.5 billion in 2012 alone. Meanwhile, unemployment in his country has reached 55 percent. United Kingdom union leader Frances O'Grady says it is essential to remember the fact that rising pay inequality was a major cause of the financial crash of 2008. "Faced with flat wages, many people borrowed to maintain their living standards whilst the very wealthy put their cash into ever more risky investments to squeeze out returns. Unless wealth is spread more broadly, we will be unable to build a sustainable recovery, as consumer spending will continue to flat-line," she says.[1]

American economist and Nobel Prize–winner Joseph Stiglitz warns that such concentration of wealth and power gives this elite a great deal of clout in setting the rules for the global economy. In areas of trade, finance, investment, and taxation they influence governments to orient policy in a way that benefits their interests, the most spectacular example of which was the bailout of the banks and other financial institutions after the financial crisis, for which they were largely responsible. In a *Vanity Fair* essay called "Of the 1%, by the 1%, for the 1%," Stiglitz writes that this elite club does not want government to invest in infrastructure, as its members gain no benefit personally from such commons as parks, public water systems, education, personal security, or medical care, things they can buy for themselves. One need only look at the state of airports, highways, railroads, and bridges in the United States to see the result.[2]

The extremely wealthy not only lobby for personal and corporate tax reductions, they hide massive amounts of money in tax havens, robbing governments of trillions of dollars in funds that could be put to the public good. Tax campaigner James Henry, a former economist with the global consultancy firm McKinsey, estimates that between $21 trillion and $32 trillion of the world's wealth has been stashed, tax-free, in offshore investments. About half of this sum is controlled by the world's richest 91,000 people, adds the *Guardian*'s Simon Bowers.

A ground-breaking 2013 investigative study of offshore tax havens by the International Consortium of Investigative Journalists (ICIJ) uncovered the names of tens of thousands of the world's wealthy who hide their money rather than pay their taxes. "Secrecy for Sale" found that the mega-rich use complex offshore structures to own mansions, yachts, art masterpieces, and other assets, and that many of the world's top banks aggressively work to provide their customers with secrecy-soaked companies in the British Virgin Islands and other offshore hideaways.[3]

The rules of economic globalization benefit the rich by encouraging competition among countries for business, driving down taxes on corporations, weakening health and environmental protections, and undermining the rights of workers. The one percent does not favour the kind of government that would protect the world's watersheds or repair and upgrade the world's aging water infrastructure, which is in desperate need of trillions of dollars of investment. After all, strong governments might bring the rule of law to the one percent and break up the club.

CORPORATIONS RULE THE WORLD

Increasingly we live in a world ruled by corporations. Today a handful of corporations control most trade in goods and services, says the UN, and many are bigger than governments. Of the top 150 economic entities in the world, 60 percent are corporations, reports the online information journal *Global Trends*. With more than 2 million employees, Walmart's annual revenue exceeds the GDP of 171 countries, making it the twenty-fifth largest economic entity in the world. General Electric is bigger than Denmark. JPMorgan Chase and Ford are both bigger than New Zealand.

Shell has larger revenues than the combined GDPs of Pakistan and Bangladesh, the sixth and seventh most populous countries in the world. Sinopec, China's leading energy and chemical company, is bigger than Singapore. Exxon Mobil is bigger than Sweden. The five largest energy corporations control the equivalent of 2.5 percent of the global GDP.[4]

However, it is not just size that gives these corporations influence. It is also how they are linked in a web of global power that is deeply disturbing for the future of democracy and the commons. An in-depth analysis by systems theorists at the Swiss Federal Institute of Technology shows that the world's transnational corporations and global banks are so interlinked that a relatively small group of companies has unprecedented control over the global economy. As *New Scientist* reports, the study revealed a core of 1,318 companies with interlocking ownerships and ties to two or more other companies; on average they were each connected to twenty. What's more, although they represent 20 percent of global operating revenues, through their shares the 1,318 appear to collectively own the majority of the world's large blue-chip and manufacturing firms — the "real" economy — representing a further 60 percent of global revenues.

When the team further untangled the web of ownership, says *New Scientist*, it found that much of it tracked back to a "super-entity" of 147 even more tightly knit companies — all owned by other members of the super-entity — that control 40 percent of the total wealth in the network. "In effect, less than 1 percent of the companies were able to control 40 percent of the entire network," says James Glattfelder, one of the authors. Most are financial institutions: the top twenty include Barclays Bank, JPMorgan Chase and Company, and the Goldman Sachs Group.[5]

The connection between big banks and big oil is startlingly clear in the report, as is the interlinking of financial investment and all the extractive industries. In an annex, the report lists fifty interconnected "knife-edge" companies (all but two are banks or finance and insurance companies) whose individual or collective collapse could lead to a far greater crisis than the one of 2008, as their fall would precipitate a catastrophic domino effect. These corporations and banks can truly be said to rule the world, and their vision is not one that respects common assets, social security, or local community control of resources.

The rise of the transnational corporations has had a huge and negative impact on the notion of the commons. No longer tied by loyalty to their home countries and not bound by international laws that might rein them in, many transnationals foul the land, air, and water of communities and use their power to undermine or avoid government regulation altogether. In many countries they obtain favourable conditions for investment such as twenty-five-year tax breaks, the right to bring in their own workers, the right to forcibly remove local people from the land, and even the right to use private security forces to "protect" their property — mines, energy pipelines, biofuel plantations — from local resisters. Transnational corporations often override democracy at all levels of government. The very thing we most need if we are to protect water — engaged local communities armed with the right to oversee and protect local water sources — is undermined by these powerful interests that write the rules to promote their profits.

Corporations also use their clout to influence elections. In the United States, corporations first gained "personhood" in a Supreme Court decision of 1886. In 2010 another U.S. Supreme Court decision removed federal rules limiting the amount of money that corporations can contribute to political campaigns, essentially preventing any efforts by future governments or states to restrict the power of big money during elections. An unprecedented $6 billion was spent on the 2012 presidential campaign, with outside groups pouring $1.43 billion into the race. Over $800 million was spent by outside interest groups in election advertising alone, most of it by corporations.

Some argue that this money was wasted, as the one with less to spend (but not by much) — Barack Obama — won. This misses the point, says Albert Hunt, former executive editor of *Bloomberg News,* who argues that the "corrosive corruption of big money" weakens political parties as well as the democratic process. The money corrupts parties because it influences the issues they choose to campaign on and the policies they adopt after winning. The biggest contributor to the Republicans was Las Vegas casino magnate Sheldon Adelson, who gave more than $90 million to the party during the election; his company is under federal investigation for possible violations of the law. (He was in Washington after the election to meet Republican members of Congress, possibly, it was reported, to discuss changes to the current anti-bribery laws.) The candidates also had to spend a huge amount of time fundraising in the "opulent homes of donors, filled with priceless art and antiques," instead of with the voters, says Hunt.[6]

CORPORATIONS WRITE THE RULES FOR TRADE

Over the past twenty-five years, corporations have been the driving force behind global, regional, and bilateral trade and investment agreements that favour their interests by limiting the ability of signatory countries to set conditions on global trade and investment. The goal of free-trade agreements is the elimination of tariff and non-tariff barriers to free movement of goods and services. Non-tariff barriers include local economic development programs, domestic food sovereignty rules, and environmental laws that are thought to be "excessive" and to hinder trade. The World Trade Organization was created to monitor the movement of goods and capital across borders and ensure compliance with strict pro-corporate rules.

Investment treaties give foreign corporations "investor-state" rights, allowing them to bypass their own governments and directly sue the government of another country if they believe their "right to profit" has been affected by a law or practice in that country. Investor-state rights first appeared in the 1994 North American Free Trade Agreement (NAFTA) and have exploded since then. There are almost three thousand bilateral deals between governments, most giving corporations these extraordinary rights, and many of them are used to gain access to the commons resources of other countries, placing the world's forests, fish, minerals, land, air, and water supplies under direct control of transnational corporations.

Canada's freshwater heritage, for instance, has been directly affected by Chapter 11, the investor-state clause of NAFTA, which allows American corporations operating in Canada to sue for financial compensation if any changes are made to the policies or practices under which they first invested. In 2002, S.D. Myers, an American company specializing in disposal of hazardous waste, including PCBS, was awarded more than $8 million from the Canadian government for loss of profit after Canada banned trade in PCBS to protect the environment and human health. Currently Lone Pine Resources, an American energy company, is suing the government of Canada for $250,000 because in 2011 the province of Quebec passed a moratorium on shale-gas fracking in order to protect its water reserves.

If the government of Alberta were ever to limit the current water access of energy companies operating in the tar sands, say legal experts, the American companies could sue for huge sums of compensation from the government of Canada. Alberta lawyers Joseph Cumming and Robert Froehlich warn that cancelling or limiting water licences would be seen as a form of trade-illegal expropriation, costing the Canadian taxpayer potentially billions of dollars. Equally worrisome, they say, is that the threat of such compensation might prevent the Alberta government from taking such a step in the first place, allowing American energy corporations to dictate Canadian policy.[7]

In a particularly disturbing development, the government of Canada awarded an American company compensation for the water rights it was no longer using after it abandoned its Canadian operation. After running a pulp and paper mill in Newfoundland for more than a century, U.S. forestry giant Abitibi Bowater declared bankruptcy and left the province in 2008. The Newfoundland government expropriated the company's assets in the province, including its water rights, in order to help pay for environmental cleanup and pensions for laid-off workers. The Newfoundland government argued that the water belonged to the province and was allocated to the company only as long as it operated a mill there. Abitibi Bowater sued the Canadian government under Chapter 11 of NAFTA, and the Harper government settled without going to a NAFTA tribunal, giving the company $130 million in compensation. This has set a dangerous precedent whereby corporations from one country operating in another can now claim ownership of local water supplies, thus providing one more way in which the world's water is becoming commodified and privatized.

Yet in spite of the profoundly undemocratic nature of the notion that corporations can hold foreign countries hostage in this way, both investor-state treaties and disputes are exploding in number. A report by the *South-North Development Monitor* on the rise of international investment disputes found that there were sixty-two new cases of corporations challenging governments for compensation in 2012, the highest number of known treaty-based cases ever filed in one year. This brings the overall number of known cases to 518 (since most arbitration forums do not maintain a public registry of claims, the total number is likely much higher, says the report). A strong majority of cases are laid by corporations from wealthy countries against countries from the developing world. This clearly demonstrates that the process works to favour powerful corporations and countries. As well, a growing number of disputes are challenging environmental rules around the world, a dangerous development that threatens the rights of governments to protect vital water sources from corporate control.[8]

Meanwhile, an elite coterie of lawyers, arbitrators, and financial speculators is making a killing seeking out and actively recruiting corporations to sue governments around the world over new health and safety, labour, or environmental rules they may be considering.

In their 2012 report *Profiting from Injustice,* Corporate Europe Observatory and the Transnational Institute say that the silent rise of a powerful international investment regime has ensnared hundreds of countries and put corporate profits before human rights and the environment. This "investment arbitration boom" is costing taxpayers billions of dollars and preventing legislation in the public interest.

Just fifteen arbitrators, all from Europe, Canada, and the United States (who can earn as much as $1 million per case), have decided 55 percent of all the treaty disputes. "They have built a multi-million-dollar, self-serving industry, dominated by a narrow exclusive elite of law firms and lawyers whose interconnectedness and multiple financial interests raise serious concerns about their commitment to deliver fair and independent judgements," say authors Pia Eberhardt and Cecilia Olivet.[9]

Undeterred, the Canadian government is deep in negotiations with Europe to seal a new form of trade and investment treaty that for the first time includes sub-national governments. The Canada-EU Comprehensive Economic and Trade Agreement will give French utility giants Suez and Veolia the right to challenge Canadian municipalities that try to remunicipalize their water services. It will also permit Swiss bottled-water giant Nestlé (whose water division headquarters are in France) to challenge provincial bans or limits imposed on bottled-water takings. A major new investment agreement with China will give the Chinese state-owned energy company, CNOOC, the right to sue the Canadian government if British Columbia forbids the building of a controversial pipeline to carry Alberta tar-sands bitumen to the west coast for tanker export. The company will also have the same NAFTA rights that American energy companies now have to fight any move by Alberta to conserve and protect its water.

CARETAKERS OF THE LAND AND WATER ARE
BANISHED

Free-trade zones are a direct outgrowth of free-trade agreements and a more recent form of enclosure of the commons. Free-trade zones, sometimes called export-processing zones, are industrial parks where foreign companies can import raw materials and export man-ufactured goods without going through customs. Trade barriers such as tariffs and taxes are usually eliminated, and so too are the labour and environmental laws of the host country. Most of these zones are located in the Global South to take advantage of cheap labour and give foreign transnational companies advantages not open to domes-tic companies. And all of them displace local people and privatize their local commons.

In his book *Development and Dispossession,* University of Florida professor emeritus of anthropology Anthony Oliver-Smith reports that capital-intensive high-technology, large-scale economic zones cause the displacement and resettlement of an estimated 15 million people every year. Farmlands, fishing grounds, forests, and villages are converted into reservoirs, irrigation systems, mines, plantations, colonization projects, highways, urban renewal zones, industrial complexes, and tourist resorts. The local people are left permanently displaced, disempowered, and destitute by what Oliver-Smith calls "development disasters." There is no return to land submerged under a dam-created lake or to a village buried under a stadium or thruway, says Oliver-Smith, and no return to the displaced rights of the com-munity and their common resources.[10]

Three thousand manufacturing operations make up the infamous maquiladora free-trade zone of Mexico, where border rivers such as the New River are so polluted that Mexican men trying to illegally enter the United States wear plastic bags on their feet to protect them from the toxins and sewage. One of the core tenets of NAFTA, the

free-trade agreement that sparked the huge growth in the maquila-
doras, was removal of the constitutional right to communal land and
water for traditional communities and indigenous peoples.

During a farmers' protest at the fifth ministerial meeting of the
World Trade Organization (WTO) in September 2003, South Korean
farm activist Lee Kyung Hae climbed a metal barricade and stabbed
himself to death in front of thousands. As with so many other farmers in his country, a free-trade zone had displaced his farm. He had
spent years counselling the families of similarly displaced farmers
who had killed themselves in despair. Just before he died, he issued
a statement: "My warning goes out to all citizens that human beings
are in an endangered situation. That uncontrolled multinational
corporations and a small number of big WTO Members are leading an undesirable globalization that is inhumane, environmentally
degrading, farmer-killing, and undemocratic. It should be stopped
immediately."

In India they are called special economic zones (SEZs); there are
hundreds of them now in operation and hundreds more planned. At
the behest of investment capital, the government buys up both farmland and villages, forcibly evicts the population, and sets up private
industrial townships for development that cover thousands of hectares of land. The zones are given tax breaks, roads, cheap electricity,
and free groundwater. They also function as self-governing autono-
mous bodies, displacing local democratic councils. A development
commissioner, appointed by government, has full powers over infrastructure decisions, access to the industrial parks, and workers'
rights.

Free-trade zones have been met with stiff resistance in India,
and the government has used force to quell dissent. In 2007 more
than three thousand heavily armed police stormed the West Bengal
village of Nandigram to remove local protesters who opposed expropriation of ten thousand acres of land to set up a chemical-hub SEZ.

Fourteen villagers were killed and at least seventy wounded in the attack. Yet protests around the country continue, as this development is destroying much of India's arable land and the communities that derive their living and culture from the land.

In China, millions of small farmers and peasants are forcibly cleared from the land to make way for developers. *Wall Street Journal*'s Tom Orlik has exposed the terrible practice whereby local governments pay small farmers just 9 yuan ($1.45) a square metre for their confiscated land and resell the land for 640 yuan to a developer, who then builds luxury villas that sell for 6,900 yuan a square metre. Researchers at the Chinese Academy of Social Science estimate that as many as 50 million landless farmers have now been forced to move to the slums ringing the big cities, where they are ineligible for social benefits because they are not registered urban residents.

China's breakneck investment frenzy (real estate investment accounted for 5.7 percent of China's GDP in 2001 but grew to just under 14 percent in 2012) is paving over food-producing land, fouling the air, and destroying local water sources. And the very people who could have saved this water — the rural small farmers and peasants — have been banished.[11] The tragic irony is that China has seriously overestimated the urban development it needs, building dozens of huge "ghost towns" — urban centres where millions of apartments and offices sit vacant and malls shuttered.[12]

LAND AND WATER GRABBED

This "depeasantization" of land and water is also escalating through the practice of what is known as "land grabbing." Wealthy countries and international investors are buying up massive parcels of land in Asia, Latin America, and Africa either to feed their own populations or as speculative investments. Using reports from the Land

Matrix Project, a global network of forty-five research and civil society organizations, the Worldwatch Institute estimates that at least 70 million hectares of land — an area almost triple the size of the United Kingdom — has now been acquired or is under negotiation between poor countries and wealthy buyers. These buyers include hedge funds, investment banks, agribusiness interests, commodity traders, pension funds, and countries such as Saudi Arabia, Japan, Qatar, South Korea, and the United Arab Emirates. Populous China and India are major players as well. This trend, which many consider to be a dangerous new form of colonial conquest, was triggered by the 2008 spike in global food prices, itself caused to a great degree by financial speculation, and the global rise in demand for biofuels.[13]

Investors are getting incredible deals, some leasing vast tracts of land for ninety-nine years for as little as forty cents per acre per year. The size of some of the land purchases is breathtaking, as are the numbers of people being displaced. State authorities in Cambodia have granted private firms concessions that amount to 22 percent of the country's land mass. Since 2003 more than 400,000 Cambodians have been affected by seizure of their land, creating an underclass of peasants with no means of income. In January 2013, armed police moved into a Phnom Penh neighbourhood and used live bullets and tear gas to quell a protest against the demolition of the homes of three hundred families to make way for a land grab. The houses were crushed before the families had a chance to collect their belongings.[14]

However, it is Africa that is bearing the brunt of these grabs. Almost 5 percent of Africa's agricultural land has been sold or leased to foreign interests. Ethiopia is forcing tens of thousands of its poorest people from fertile land earmarked for Saudi and Indian investors, reports the *Telegraph*'s Mike Pflanz. Already twenty thousand families have moved from areas they have roamed for centuries so that private companies can take over 400,000 hectares of land in the Gambela region alone. A further 2.4 million hectares of Ethiopia, an area almost the size of Belgium, have been leased to foreign companies, and there are plans to expand this expropriation by another 2 million hectares. Human Rights Watch says the evictions have been accompanied by widespread human rights violations, including forced displacements, arbitrary arrest and detention, beatings, and rape and other sexual violence.[15]

In many countries these smallholder farmers provide the majority of food for the local population. With them gone and their land used to grow crops for export, who will feed local communities? Replacing local biodiverse farming with industrial agribusiness removes large tracts of land from domestic food production and replaces it with chemical-fed monoculture crops for export. John Vidal writes in the *Guardian* that Ethiopia is one of the hungriest countries in the world, with some 2.8 million people needing food aid. Paradoxically, its government is offering up vast tracts of its most fertile land to rich countries and rich foreign investors to grow food for export.[16]

Land grabs are also water grabs. Investors intending to set up intensive commodity production need guaranteed access to plentiful water, and that means irrigation using surface and groundwater supplies. Not only do these corporate farms choose the best land for their crops, they tie up access to the water rights of local streams, rivers, and aquifers. And they often get the water for free or pay a pittance for it. Even Nestlé executive and World Bank advisor Peter Brabeck admits that the global rush for land is really a water grab: "with the land comes the right to withdraw the water linked to it, in most countries essentially a freebie that increasingly could be the most valuable part of the deal."[17]

The impact on local watersheds could be truly devastating. These deals and the land sold as a result are replicating the worst models of corporate farming from the industrial world, complete with water pollution, over-extraction, flood irrigation, and the export of local water sources through the consequent global trade in the commodities they produce.

Food and Water Watch reports that some of these foreign-controlled industrial farms are so big it would be like foreign investors buying up and owning all the farmland in Oklahoma or North Dakota. GRAIN, an international non-profit that promotes community-owned food systems, reports that the land already leased in Sudan, South Sudan, and Ethiopia alone, if fully cultivated, would require much more water than is available in the entire Nile Basin, amounting to "hydrological suicide."[18]

The San Francisco–based Oakland Institute says the volume of water required in Africa to cultivate the 40 million hectares of land acquired in 2009 alone will require 300 to 500 cubic kilometres of water per year, twice the volume of water that was used for agriculture in all of Africa in 2005. If the rate of land acquisition continues to grow apace, says the institute, demand for fresh water will overtake the existing supply of renewable water on the continent by 2019. This will jeopardize Africa's fragile river systems and divert even more water from rivers and lakes already under stress, such as the Niger River.[19]

Journalist Claire Provost, writing in the *Guardian*, says it is no coincidence that the most aggressive foreign land investors are also those facing water shortages at home. Water is the reason that India, South Korea, and China are racing to buy land and grow crops in other countries, she says, pointing to water-stress indexes that indicate crises in all those countries. Saudi Arabia has actually established a new agricultural fund whose prime concern is preserving domestic water supplies by investing in agricultural production in other countries. Provost quotes a report by James Skinner and Lorenzo Cotula, of the London-based International Institute for Environment and Development, that says an alarming number of African countries are signing away water rights for decades, in many cases free of charge.

Skinner and Cotula report that long-term contractual commitments with investors can jeopardize water access not only for those living near the industrial farms but also for those living downstream. In some cases, they say, estimates of the potential water requirements are so large that major dam projects are being considered to ensure supply. The controversial Gibe III Dam in Ethiopia will help irrigate 150,000 hectares that the government has leased to foreign investors. The volume required could seriously lower the water level of Kenya's Lake Turkana.[20]

Not surprisingly, the same financial and trade institutions that have promoted water privatization in the Global South are funding and protecting this latest enclosure of the commons. Shepard Daniel and Anuradha Mittal of the Oakland Institute report that the International Finance Corporation (IFC) of the World Bank officially endorses an increase in global agricultural production; it works with and heavily funds agribusiness and the agro-industry in "emerging market countries" in order to promote the role of the private sector. The IFC believes that high food prices offer unique opportunities for emerging markets to grow and develop their agricultural sectors, based on an industrialized model. The IFC works with governments in poor countries to reduce their demands that foreign companies leave some profit in the country and to change their land laws to increase the permissible quantity of land under foreign ownership.[21]

Not surprisingly, trade and investment treaties create an additional set of rights for the foreign investors in these deals. Dr. Howard Mann and Carin Smaller, of Canada's International Institute for Sustainable Development, write that these new land grabs differ from the traditional form of foreign investment in domestic food production in that it is no longer just the crops that are commodities. Rather, it is the land and water themselves that are increasingly becoming commodified and therefore increasingly subject to global rules of rights of access. The growth of investments in the actual land and water, not just the crops, increases the potential to shift rights from domestic to foreign players, they say. Most of the host countries do not have domestic laws that adequately protect their land rights, water rights, resources, workers, or small farmers.

"On the other hand, the international law framework provides hard rights for foreign investors," say Mann and Smaller. This layering of international law over domestic law has "potentially disastrous legal consequences," such as huge compensation claims if the host country decides to reclaim some of the leased or sold land and water for environmental purposes or for its own people.[22] In other words, once established, these unsustainable mega-farms will very likely be entrenched and protected in international trade law.

GLOBAL FRAMEWORK FOR WATER COLLAPSE

Clearly economic globalization, with its emphasis on growth at all costs, its servitude to the one percent, its systematic enclosure of the commons, its entrenchment of corporate rights in international trade law, and its displacement of the local caretakers of land and water everywhere, is a sure-fire recipe for water disaster. This is truly "modern water" come upon us. The solution to the global water crisis must include renunciation of this model of growth if there is any hope that it will be successful. Trade must be radically reformed to serve a different set of goals and come under democratic oversight. Corporations must lose the right to sue governments. Land grabs must end. Tax havens must be shut down. The rule of law must be brought to bear on transnational capital. Citizen-driven democracy must be restored or, if necessary, built from the ground up. To do this, we must overcome our political apathy.

The antidote to bad governance is not transnational corporations running the world in their image. The antidote to bad governance is good governance. Only the power of true democracy will bring about the conditions necessary to protect the world's water.

14

CREATING A
JUST ECONOMY

Our vision is of a just global economy where humans are treated
justly and with dignity, where the environment is respected and
nourished, where commerce fosters sustainable communities and a
global society based upon cooperation and solidarity. —Fair World
Project (Portland, Oregon), "System Change for a Better World"

TRADING FOR PEOPLE AND THE PLANET

In 1993, Jerry Mander, a writer, activist, and organizer from San
Francisco, called a gathering of some of the most remarkable people on the planet to build a critique of and resistance to economic
globalization. Most of those who came together were veterans of
trade campaigns such as NAFTA and the General Agreement on
Tariffs and Trade (GATT), and all of them were preparing to fight
the World Trade Organization, which would be formally established
two years later. Out of this gathering came the International Forum
on Globalization, whose core members were writers, thinkers, and
environmental and activist leaders from all corners of the globe. At
Jerry's suggestion, we revived the 1960s notion of the "teach-in" and

held large-scale public events at many major gatherings, including the WTO ministerial "Battle in Seattle" in 1999 and the Rio+10 Earth Summit in Johannesburg in 2002. Many of us went on to form Our World Is Not For Sale, an international network of organizations and communities fighting regional free-trade agreements and coordinating opposition to the WTO.

We were clear from the first that we were not opposed to trade or to trade rules but that we deeply opposed the kind of trade and investment agreements that put corporate profits ahead of people and the environment. As Stuart Trew, trade-justice campaigner for the Council of Canadians, explains, these free-trade deals purport to promote a twenty-first-century model of investment but only reproduce an exploitative late-nineteenth-century relationship between global capital, workers, and the earth. Our core belief is still that trade and the economy should serve people and communities, not the other way around. We were, of course, accused of being narrow "protectionists," as if wanting to protect the environment, the health and safety of families, or the rights of workers from exploitation were a negative thing.

Pro-globalization advocates promised that the drive to privatize public assets and free the market from governmental interference would spread freedom and prosperity around the world, improving the lives of people everywhere and creating the financial and material wealth to end poverty and protect the environment. Two decades later, this promise has been exposed for the lie that it was. We stand at the brink of unparalleled species extinction and a growing gap between those who have benefited from globalization and its victims. An April 2013 report on globalization by the OECD says that, while there are pockets of positive stories overall, the rich have got richer and the poor poorer after thirty years of this experiment. "The poorest country in 2011 was poorer than the poorest country in 1980. And much of humankind continues to live on less than USD$1 a day.... In many countries, inequalities have deepened."[1]

As stateless corporations gave rise to corporate states, we built a global movement for trade justice to fight a plethora of new trade deals. We were successful in stopping the Free Trade Area of the Americas, a proposed expansion of NAFTA to all of Latin America. We brought the WTO to its knees in Seattle and, working with nation-states from the Global South, we have succeeded in stalling the WTO for two decades.

We came together in the late 1990s to defeat the Multilateral Agreement on Investment (MAI), which would have given global corporations NAFTA-type investor-state rights around the globe. Canada's *Globe and Mail* said that the "high-powered politicians" around the world were no match for "a global band of grassroots organizations," and that defeat of the MAI had "transformed" international relations. The *Financial Times* compared the fear and bewilderment that seized the governments of the industrialized world to a scene from the movie *Butch Cassidy and the Sundance Kid*: the politicians and diplomats looked behind them at the "horde of vigilantes whose motives and methods are only dimly understood in most national capitals," and asked, "Who are these guys?"[2]

A number of countries are taking steps towards a different trade model. Launched in 2004, the Bolivarian Alliance for the Peoples of Our America (ALBA) is a co-operation agreement whose partner states include Bolivia, Venezuela, Cuba, Ecuador, Nicaragua, Antigua and Barbuda, Dominica, and Saint Vincent and the Grenadines. The bloc promotes a different kind of economic integration based on solidarity, equality, justice, and integration, "where the returns on social investment are not measured in monetary terms but rather in improving peoples' wellbeing," in the words of Venezuelan trade union leader Ruben Pereira.[3]

These and a number of other countries, notably Guatemala, El Salvador, Honduras, and Mexico, met in April 2013 to form an alliance to fight back against the increasing number of lawsuits taken against

them by transnational companies under investment treaties. The
countries agreed to establish a permanent conference of states to deal
with the challenges posed by the power of transnational corporations
to sue foreign governments. Australia has banned the negotiation of
trade deals that include any type of investor-state clauses, and Brazil,
which now has the tenth largest GDP in the world, is not a party to
any bilateral investment treaties and has not ratified the International
Centre for Settlement of Investment Disputes Convention.

Australia and Brazil must become a model for every country
in the world. Investor-state clauses that give corporations the right
to sue foreign governments for compensation or to place a chill on
governments considering new laws and practices to protect their
environment, the health and safety of their people, or social rights
must go. Thomas McDonagh of the San Francisco–based Democracy
Center, says that investors can be legally protected in international
agreements, but that these agreements would have to adhere to certain principles. These include

- putting human rights before corporate rights;

- creation of a new system of dispute resolution that includes
  domestic courts;

- binding obligations on corporations;

- policy space for local economic development;

- capital controls to stem financial speculation; and

- restrictions on the definition of *investment* to prevent
  investors from interfering in a country's right to set environmental and social standards.[4]

In *Alternatives to Economic Globalization*, edited by Jerry
Mander and John Cavanagh, the brilliant executive director of the

Washington-based Institute for Policy Studies, we laid out the basic principles of an alternative model of trade and development:

- All trade and economic activity must promote local democracy and seek to devolve power to those who will bear the costs of decisions made higher up.

- Promoting the concept of subsidiarity, which gives priority support to goods and foods that can be produced locally, all rules of trade and investment must recognize that food production for local communities should be at the top of a hierarchy of values in agriculture, and that local self-reliance in food production and assurance of healthful, safe foods should be considered basic human rights.

- All economic and trade activity must be ecologically sustainable, taking care to ensure that rates of renewable resource use do not exceed their rate of regeneration and that rates of pollution emissions do not exceed the rate of their harmless assimilation.

- All economic and trade activity must recognize and promote cultural diversity, which is key to the survival of indigenous peoples; economic diversity, which is the foundation of resilient, stable, energy-efficient, self-reliant local economies; and biological diversity, which is essential to complex self-regulating, self-regenerating processes of the ecosystem.

- All economic development and trade activity and policy should enhance the core labour rights and human rights included in the Universal Declaration of Human Rights, and the two covenants ensuring economic, social, and cultural rights as well. This requires protection in law of the rights not only of workers but also of people who subsist in the informal work sector, including subsistence farmers and those who are unemployed or underemployed.

- All the global governance institutions must address the growing gap between rich and poor — between nations and within nations.

- And finally, all economic and trade activity must promote the precautionary principle, putting the onus on the producer to prove the safety of a product, as opposed to the system of risk assessment under current trade rules, in which the onus falls on governments to prove harm.[5]

Walden Bello, one of the International Forum on Globalization founders and a member of the Philippine House of Representatives, says that globalization has been "terminally discredited" in the past several years and that it is time to recognize the end of an era. In his call for "deglobalization" he would use trade policy to protect local economies, implement long-postponed measures for equitable income and land redistribution, de-emphasize growth while promoting quality of life, expand the scope of democratic decision making so that all vital questions become subject to democratic debate, and replace the International Monetary Fund and the World Bank with regional institutions based not on free trade and capital mobility but on the principles of co-operation.[6]

To counter the kind of trade supported by most governments and corporations, the fair-trade movement was created to help producers in the Global South find markets for products made under good working and environmental conditions. It focuses on handicrafts and commodities such as coffee, cocoa, sugar, tea, bananas, honey, and cotton, which must be grown and harvested in accordance with international standards.

There are several recognized fair-trade certification systems. The largest, Fairtrade International (FLO), handles more than a thousand certified companies and has sales of more than $5 billion a year. With coffee, for example, packers in developed countries pay a fee to the Fairtrade Foundation for the right to use the brand and logo. The coffee has to come from a certified fair-trade co-operative, which pays certification and inspection fees. The importer pays the exporter more than the going world price in order to guarantee a living wage to the farmer, knowing that there is a market for food grown under better conditions. By all accounts, the future for fair-trade products is excellent.

TRADE THAT PROTECTS WATER

Given the threat to water from existing and proposed trade and investment agreements, it is urgent to remove all references to water as a service, good, or investment in all present and future treaties. Water is not like anything else on earth. There is no substitute for it, and we and the planet cannot survive without it. Water must not be a tradable good, service, or investment in any treaty between governments, and corporations should have no right to stop domestic or international protection of water.

Further, trade negotiations should take into account the effect on water of all trade activities. Arjen Hoekstra, the virtual-water expert, says officials should ask (and be able to answer) these questions:

- how much water was consumed to make a product in the different stages of its supply chain;

- how much water was polluted and with what type of pollution;

- whether the water consumption took place in areas where water is scarce or abundant;

- whether the ecosystem or downstream users were affected by the water consumption; and

- whether the water consumed could have been used for an alternative purpose with a higher social benefit.

Hoekstra says that two cotton shirts may look identical but have totally different water footprints, depending on where the cotton was produced. Cotton from Uzbekistan and Pakistan, for example, can be directly associated with desiccation of the Aral Sea and pollution of the Indus River, respectively.

Under the "non-discrimination" principle of trade agreements, however, consideration of the origin of a product and its possible negative impact on local water supplies is not currently permitted. Furthermore, because there are no internationally binding agreements on sustainable use of water in the production of goods and services, trade disputes with respect to freshwater protection are settled by the WTO under rules that do not allow domestic environmental laws to affect the outcome.

But governments should have the right to ban products that harm water in their own countries. "Fair international trade rules should include a provision that enables consumers, through their government, to raise trade barriers against products that are considered unsustainable, or...responsible for harmful effects on water systems and indirectly on the ecosystems or communities that depend on those water systems," Hoekstra writes in a paper for the World Trade Organization. Thus one country might favour imports of a product from another that could guarantee that the product's water footprint is not located in catchments where environmental flow requirements are violated or where ambient water quality standards are not met.

Hoekstra also recommends an international label for water-intensive products that would indicate whether the product met a certain set of sustainability criteria.

As well, because water is usually given away to export-oriented agribusiness and often even subsidized by governments, water is not considered a factor of importance in the establishment of production and trade patterns. This allows water-intensive crops to be exported on a large scale from areas where water is scarce and overexploited, says Hoekstra. He calls for an international water-pricing protocol to end the market distortion of free water; he says this pricing would force companies to take into account their impact on local water sources and question the economic feasibility of inter-basin transfers of water.[7]

In the wrong hands, I fear such a protocol would simply allow those with enough money to pay to deplete or pollute water and get around the intent of the instrument. However, it is true that large users of raw water should not be getting it for nothing. It is time to start charging large-scale for-profit industries for access to raw water, as long as this access meets environmental and sustainability standards set by the community. A combination of domestic bans or restrictions on water takings for export in water-scarce areas; establishing licence fees for sustainable domestic takings of raw water; a system to ban imports that harm the ecosystems and watersheds of the country of origin; and removing water as a tradable good, service, or investment from all trade and investment treaties would provide a better framework to protect water in international trade.

REINING IN RUNAWAY SPECULATION

Equally important is the need to challenge the power of the central bankers and financial speculators who currently operate largely outside the reach of democratic control at both national and international levels. Financial speculation allows investments not in tangible things such as goods and services but in risky financial transactions that attempt to profit from fluctuations in the market value of assets. This is called "financialization of the economy," in which trading money and risk has become more profitable than producing goods or providing services.

Europeans for Financial Reform, a coalition of labour unions, civil society organizations, and the Global Progressive Forum dedicated to reform of the financial and banking sectors, launched a campaign called Regulate Global Finance Now! in 2009, the year after the financial collapse. They called on governments to regulate speculative funds such as hedge funds and private equity funds, create a tax on financial transactions (sometimes called the "Robin Hood tax," as it would tax the rich), limit executive and shareholder bonuses and remunerations, close down tax havens altogether, protect consumers from toxic financial and predatory lending, and democratize finance.

The financial crisis revealed a dangerous "shadow banking system," says the coalition, that permitted the buildup of excessive debt and speculative financial gambling. This in turn had devastating consequences for jobs, wages, the environment, and the well-being of the planet. The crisis must be a trigger for reform of the global economic order, say the groups involved, and usher in a new paradigm that privileges sustainable development and social justice.[8] Similar calls for reforms are taking places in countries around the world.

In at least one area they have had some effect. In an astonishing turnaround, many European and British banks have stopped speculating in food prices following scathing reports by Oxfam France, Oxfam Germany, Oxfam Belgium, and the World Development Movement in the United Kingdom that they were making money from hunger and poverty. The year 2008 saw near record-breaking increases in both real food prices and prices in the commodity futures market. Barclays Capital, the investment arm of Britain's Barclays Bank, made more than £500 million betting on food staples in 2010; meanwhile, South Africans faced a 27 percent hike in wheat prices, and maize prices soared by 174 percent in Malawi.[9]

The *Financial Times* reports that accusations of profiting from the global food crisis transmitted "high-voltage shocks across the fund management industry" and that the change in their conduct has been "dramatic." BNP Paribas, France's largest provider of agricultural commodities funds, suspended a $214-million agriculture fund. Landesbank Berlin, Landesbank Baden-Württemberg, and Commerzbank, in Germany, and Österreichische Volksbanken, in Austria, all pared down their exposure to the food sector. Barclays announced that it would cease the trade in food, as "this activity is not compatible with our purpose."[10]

RESISTING THE FINANCIALIZATION OF NATURE

The next step in currency speculation is financialization of the natural commons, including water, taking advantage of a process that was started for other reasons. For decades, environmentalists and scientists have been touting the virtues of what they call "ecosystem services" — the many benefits that humans receive from nature — in an attempt to get governments to enact environmental protection. In 2001 the UN initiated the Millennium Ecosystem Assessment, a four-year study involving more than 1,300 scientists worldwide. The scientists grouped ecosystem services into four categories in an attempt to quantify nature's contribution to our well-being and health. Ecosystems provide us with food, clean air, and drinking water; they regulate climate and help control disease; they support nutrient cycles and crop pollination; and they provide us with cultural, spiritual, and recreational services as well.

Unfortunately the UN went a step further than recognizing eco-system services and decided to put a price on them. It estimates that ecosystems and the biodiversity that underpins them generate services to humanity worth as much as $72 trillion a year. The reasoning for putting a specific price on them is that, if we can prove that nature has a concrete monetary value, it can compete in its natural state with the other uses to which the land might be put. Certainly it is a good thing when public or aid funds help farmers rotate their land for conservation, help fishing communities conserve stocks, or support indigenous communities in protecting the forests they live in, as preservation of land, water, and species benefits everyone.

However, under the guise of a "green economy," many countries, as they deregulate resource protection, are redefining nature as "natural capital" and assigning concrete prices to their forests, wetlands, and waterways. And financial speculators are lining up to cash in. Antonio Tricarico, a former economic correspondent with the Italian newspaper *Il Manifesto,* now with Rome-based Re:Common, calls this the financialization of nature. Combined with growing competition at the global level for control and management of natural resources, this trend could put management of the commons into the hands of financial markets for years to come, he told a crowd at the 2012 Alternative Water Forum in Marseille.

"This approach is a long-term project that aims to lock natural resources management into the future structure of capital markets in a way that will dramatically reduce the possibilities to reclaim the commons and their collective management by affected communities," he said. "This systemic 'financial enclosure' of the commons, coupled with existing trade and investment liberalisation agreements, would produce a long-lasting legal enclosure that drastically shrinks the political space for social movements reclaiming the commons as the basis of their livelihoods."[11]

What is disturbing about the current trend is that a for-profit market is being established to "protect" the environment by the same corporations that have harmed it in the first place, and nature itself is becoming an "economy" to be managed. In this scenario, payments for ecosystem services such as water purification, crop pollination, and carbon sequestration will help the environment self-regulate and the private sector will protect what's left of nature.

The process of bringing nature under control of the "logic of the financial markets" happens in stages. First, says Food and Water Watch (FWW), it is made a commodity—the commercialization of something not generally seen as a product. Commodification turns an inherent value into a market value, enabling it to be bought and sold on a market. Privatization transfers control and management of commoditized resources from public ownership to private ownership. The commodities can then be priced and a market can be created for them. At this point, financialization acts upon the commodity as an asset and applies various financial instruments to it, such as the water futures contract of a carbon credit option.[12]

An early example of "nature trading" took place in Europe with the trade in carbon dioxide pollution permits as part of a "cap-and-trade" scheme to reduce greenhouse gas emissions. This turned CO_2 into a financial asset and introduced market volatility to the fight against global warming. As a result, European corporations made windfall profits from wildly fluctuating prices. In another example of this kind of trading, polluters get "credits" for investing in large-scale industrial tree plantations in the south. The problem is that these industrial forests displace existing sustainable community logging and loggers. Friends of the Earth International and others have condemned this practice.

Coca-Cola is promoting a concept called "water neutrality," whereby the company would "offset" its water impact in one place by investing in water-saving technologies or water-provision programs

in another place. This could easily lead to a for-profit "water neutrality" market similar to the carbon market. Water scientist Arjen
Hoekstra says that, aside from the impossibility of most water-
intensive companies ever being able to become truly water neutral,
the water market that will "undoubtedly" grow up around it could
easily end up simply raising funds for charity projects in the water
sector. These "charitable works" are very important for greening the
reputation of the bottled-water industry.[13]

Water-pollution trading has already begun in the United States.
Under such schemes, explains Food and Water Watch, water
polluters essentially enter a cap-and-trade system similar to carbon-
emissions trading, whereby companies can sell and trade the right
to pollute. Regulators allow traders to purchase pollution-reduction
offsets from other, frequently unregulated sectors such as industrial
and agribusiness runoff. FWW and Friends of the Earth U.S. have
gone to court to stop a water-pollution trading plan for cleanup of the
Chesapeake Bay watershed, warning that forty years' success of the
Clean Water Act will be traded away if polluters are able to buy their
way out of upgrading equipment and reducing their toxic waste. They
point out that the coal-fired power plant industry has been quick to
adopt the notion of water-pollution trading, as it sees this as a way
to avoid technological responses to the massive amounts of nitrogen
pollution from its operations that are killing local waterways.[14]

England now has a Natural Capital Committee and an Ecosystem
Markets Task Force that promise "substantial potential growth in
nature-related markets" in the order of billions of pounds globally
and substantial returns on investment in environmental bonds. The
Guardian's George Monbiot says that payments for ecosystems services are the prelude to the biggest privatization of the commons his
country has ever seen and that his government has already started
describing landowners as "'providers' of ecosystem services, as if they
had created the rain and the hills and the rivers and the wildlife that

inhabits them." He adds that the government is experimenting with "biodiversity offsets" by allowing, for example, a quarry company to destroy a rare meadow if by a market offset it can pay someone somewhere else to create something similar.

"We don't call it nature any more: now the proper term is 'natural capital'. Natural processes have become 'ecosystem services', as they exist only to serve us. Hills, forests and river catchments are now 'green infrastructure', while biodiversity and habitats are 'asset classes' within an 'ecosystem market'. All of them will be assigned a price, all of them will become exchangeable," he says. The financialization of nature will forestall democratic choice, Monbiot adds, as governments won't need to regulate, turning the decisions they have avoided over to the market. "It diminishes us, it diminishes nature. By turning the natural world into a subsidiary of the corporate economy, it reasserts the biblical doctrine of dominion. It slices the biosphere into component commodities...If we allow the discussion to shift from values to value — from love to greed — we cede the natural world to the forces wrecking it."[15]

SAYING NO TO WATER SPECULATION

It is essential to stop this dangerous trend — it patently isn't working. We must challenge governments to take up the responsibilities they have relinquished. Ontario organic farmer Ann Slater says payments for ecosystem services are just another way for corporations to access public money. The financialization of water is not about protecting the environment, says Food and Water Watch, but about creating new ways for the financial sector to continue to earn high profits from the water crisis. We need to tell our governments to do their job rather than speculating in water, to introduce and implement the necessary regulations to protect our common ecosystems and watersheds.

Friends of the Earth International says the UN should back off from its endorsement of this trend: "The debate on biodiversity should not be reduced to just the economic benefits brought by biodiversity. The UN needs to discuss how to strengthen local communities' and Indigenous Peoples' initiatives that have contributed to biodiversity conservation and sustainable use of biodiversity, and to the construction of a fairer and more sustainable world." Isaac Rojas, the group's forests and biodiversity coordinator, adds that market-based mechanisms and the commodification of biodiversity have failed both biodiversity and poverty alleviation.[16]

In fact, the United Nations Environment Programme did undertake an innovative study to predict which kind of development — public or private — would be best for the environment in the future. It concluded that privatization, a key touchstone of economic globalization and precursor to the financialization of nature, provides the worst scenario for protection of ecosystems, even though (or perhaps because) it assures the strongest growth. Privatization "manifests an environmental impact deemed unbearable, all while generating ever-greater social inequalities," says the report, adding that in the privatization scenario, "the environment and society rapidly reach, even cross over the tipping point."[17]

Food and Water Watch executive director Wenonah Hauter says that all the pieces of water privatization coalesce in a fight against this vision of financialization of nature. Desensitization to the commodification of water through the promotion of bottled water has opened the door to the idea of water's being just like any other commodity, she says. The privatization of municipal systems, which removes from government such a basic communal function as the distribution of clean water, has made some people ready to accept the idea of water distributed by markets. All over the world, she adds, neoliberal environmental groups are promoting market pricing of water, and — as noted above — some have gone so far as to promote trading of water-pollution credits. These are all steps towards achieving water markets. "We must act now to stop this vision from coming to fruition. If we want to keep the global water market and water-based derivatives from becoming a reality, we have to work *now* to raise awareness that this is the plan; this is what many economists — including respected economists at powerful banks — see not just as the future, but as the desirable future." Or as George Monbiot advises, "Pull up the stakes, fill in the ditch, we're being conned again."[18]

CHALLENGING THE PRIVATIZATION OF FOREIGN AID

While the issue of foreign aid for water was dealt with in an earlier section of this book, it is crucial when talking about a just economy to sound the alarm on the trend to privatize foreign aid. In 2010, external investments in the private sector by international financial institutions exceeded $40 billion. By 2015 the amount of public money flowing to the private sector is expected to surpass $100 billion, almost one-third of the total amount of aid to developing countries. This is of course dramatically increasing the power of the private sector to decide how aid money is spent and which governments will receive aid. The Canadian government, which openly provides foreign aid through its mining industry, has stated that it will take into account the foreign investment policies of the recipient countries when deciding where to put aid money, in order to ensure that their policies are compatible with the interests of Canada's mining companies.

As aid is privatized, the agencies delivering it are politicized. African water activists (who cannot allow their names to be used for fear of retaliation) say that most of the international water NGOs and aid agencies practise "selective funding," tending to either partner with local NGOs that are ideologically aligned with them or create their own surrogate NGOs to operate within their ideological ambit. Together, these international aid agencies and their local counterparts promote public–private partnerships as the model for water services. This in turn benefits the private water utilities, which often fund these aid agencies.

WaterAid, a charity started by British water companies, promotes the charity work of many corporations, including private utilities, in countries throughout Africa and Asia. This allows corporations to "blue-wash" bad practices. WaterAid recently partnered with Coca-Cola to provide clean water in a number of African countries, a project that the company, strongly criticized around the world for depleting local water supplies, has heavily advertised.

In 2012 the British multinational bank HSBC launched a $100-million "partnership" with the World Wildlife Fund, Earth Watch, and WaterAid to "bring improved water and sanitation systems to the poor in India, Pakistan and Africa." But African communities have had a very different experience with this bank. HSBC is the major investor behind a company called New Forests, which grows forests in Africa to use as offset credits for polluters abroad. Oxfam International reports that more than 22,500 people were forcibly evicted from their land and water sources in Uganda to make way for a New Forests project.[19]

While governments and international institutions are promoting the privatization of development aid, a report by the European Network on Debt and Development found that most of the recent development investments are going to private tax havens and private companies. The study analyzed more than $30 billion worth of private-sector investment in the world's poorest countries funded by "development investments" from the World Bank's International Finance Corporation, the European Investment Bank, and a number of European national development financial institutions between 2006 and 2010. Only a quarter of all recipient companies were based in low-income countries, while almost half (49 percent) were based in one of the thirty-four OECD developed countries. Forty percent of the recipient companies are listed on the largest stock exchanges and half the money was given directly to the financial sector, commercial banks, hedge funds, and private equity funds. Over a third of the money was commissioned to companies based in tax havens.

The report's author, Jeroen Kwakkenbos, notes that many see nothing wrong with this system. People working for the World Bank come from the banking sector, not the development sector, and have a finance-oriented take on things. In an interview with Inter Press Service he questioned the wisdom of public funds going to these corporations in the name of development. "What kind of development priorities are these companies looking at? How do they align themselves with country priorities in developing nations if they keep on bypassing the government and investing directly in the private sector? And, most importantly, how do you marry profits and development objectives? What is most important in the project? Is it development outcomes or the long term return on investments?"[20]

This shocking report underlines the urgent need for reform and accountability in the world of aid and development and adds another reason to support Walden Bello's call to dismantle the World Bank. Governments and international aid agencies are handing out public money. Every cent of it should go to building a just and sustainable economy for all, not simply lining the pockets of wealthy corporations.

15

PROTECTING LAND, PROTECTING WATER

With our gifts as humans to reach out, measure, hypothesize and organize our discoveries into conceptual theories we, the stuff of stardust, have the privilege to know what has come before, how we are now and where we might be going. At this point of Earth's journey, we are Earth, reflecting on herself. —**Professor Ralph C. Martin, University of Guelph, Ontario**

A WATER-SECURE WORLD IS dependent on respect for and sharing of land and watersheds around the world. Since growing food accounts for so much of the water consumed everywhere, it is essential to change the way we use land and to challenge both the corporate domination of food production and the displacement of peasants, indigenous peoples, and small farmers from their land. Ralph Martin, who founded the Organic Agriculture Centre of Canada and teaches lucky young people at the University of Guelph, says that we have to ask ourselves if current methods of food production have negative effects on earth's self-regulating biological systems, and if the answer is yes, they must be changed.

Paraphrasing several people, including Thomas Berry, Brian

Swimme, Miriam Therese MacGillis, and others, Martin asks us to remember how recently we humans and our modern world came to the universe. The universe flared forth about 13.5 billion years ago, says Martin, and earth was formed about 4.5 billion years ago. Looking at our planet's history as if it had taken place over a single year, he calls this January 1. By 3.9 billion years ago (February 18), single-cell life had developed. It wasn't until 2.5 billion years ago (June 11) that oxygen and ozone appeared, and then finally, 600 million years ago (November 12), animals graced the earth. Plants and fungi followed 440 million years ago (November 25), and then dinosaurs arrived 145 million years ago (December 19). Creatures resembling humans were latecomers 5 million years ago (December 31, 9:44 a.m.).

"Agriculture," says Martin, "so crucial to our lives, is really only a blink of Earth's story with its flirtatious entry 10,000 years ago, or one second before the end of the year. Modern agriculture has only existed for about 100 years, that is one per cent of a blink in Earth's history. Earth's self-regulating biological systems were selected over millenniums for resilience, delightful diversity and zero waste. It is in this context that we can choose how to raise food and live."[1]

PROMOTING LOCAL, ORGANIC, SUSTAINABLE FARMING

The cardinal goal for national governments and international institutions alike must be food and water security for local communities, and all policies and practices must support this objective. Further, authority for food and water must be devolved to local communities and decision makers wherever possible; governments must work to provide the political environment and legal protection for people and communities to determine their food and water priorities. Farmers have the right to a fair price for their goods and should not be discriminated against in policy decisions or used as pawns by agribusiness corporations. Trade policies must allow countries to set domestic food-production targets and conserve national supplies by stopping exports in time of scarcity.

The alternative to the current global food system, for land, for climate, for water, and for people, says Indian physicist and environmental leader Vandana Shiva, is biodiversity-based, locally run and controlled organic farming. Biodiverse, ecologically sensitive farms address the climate crisis by reducing emissions of greenhouse gases such as nitrogen oxide and absorbing carbon dioxide into plants and the soil. Biodiversity and compost-rich soil are the most effective carbon sinks. In turn, they increase organic matter, which increases the moisture-holding capacity of soil and hence provides drought-proofing of agriculture. Biodiverse, organic farms increase food security by increasing the resilience and reducing the climate vulnerability of farming systems, says Shiva, and they enhance food security because they produce a higher yield of food and nutrition per acre than chemical-based industrial food production.

Biodiverse organic farming also addresses the water crisis in three essential ways. First, production based on water-prudent crops reduces water demand. Countries with dry lands have had centuries to develop hundreds of thousands of drought-resistant crops, Shiva points out, something the biotechnology industry fails to understand in its race to produce drought-resistant corporate-controlled monoculture crops. (Jodi Koberinski of the Organic Council of Ontario says genetically modified foods are not about feeding the world, as the industry claims, but about controlling the seed.) Second, says Shiva, organic systems use ten times less water than chemical systems and don't poison local water supplies. Finally, by transforming the soil into a water reservoir by increasing its organic-matter content, biodiverse organic systems reduce irrigation demand and help conserve water in agriculture.[2]

Shiva calls on the government of India to reject "the Green Revolution model of water-intensive chemical farming," as well as foreign land grabs. She advocates that the government support an alternative that includes conservation and distribution of water-prudent crops and seeds, as well as incentives to farmers to return to biodiverse organic agriculture in order to increase climate resilience and food and water security.[3]

Sandra Postel writes that a spontaneous, largely under-the-radar "blue revolution" is gaining steam in sub-Saharan Africa that has the potential to boost food security, protect water, and provide income for millions of the region's poorest inhabitants. Small-scale irrigation techniques, using simple buckets, affordable pumps, drip lines, and other equipment, are enabling farm families to weather dry seasons, raise yields, diversify their crops, and lift themselves out of poverty. Most of these countries have barely begun to reach their food production potential. Done right, such an initiative could provide food and water for all. With access to pump-irrigation water, farm families and communities reap bigger harvests, greater food security, and more income. Conservation farming that retains rainwater in the soil can greatly improve productivity on small farms, says Postel, noting that most of the cropland in this region of Africa is watered only by rain, making rainwater harvesting a key tool to sustainable farming.[4]

The Oakland Institute gives a number of examples where a focus on efficient small-scale irrigation, sustainable agriculture, and water management methods can improve the lives of local smallholders, enhance food security, and prevent environmental degradation from water depletion. In Zimbabwe, sustainable water management and water harvesting systems have proven very effective in increasing yields, building resistance to climate shock, and improving income and food security. In Burkina Faso, the introduction of soil and water conservation techniques has led to economic security and population stability, as well as improved water tables in several areas. In Ghana, the production of drought-resistant staples such as millet and sorghum shows better yields with small-scale rather than large-scale irrigation. In Kenya, bio-intensive agriculture, a low-cost technology designed for small farmers, has been shown to use 70 to 90 percent less water than conventional agriculture.[5]

REJECTING LAND AND WATER THEFT

As both Postel and the Oakland Institute point out, unless African governments, foreign interests, and international institutions support these farmer- and community-driven initiatives, the best opportunity in decades for social advance in the region will be squandered, as will the opportunity to make the region water-secure.

As it is, most governments in Africa have not yet adequately addressed their water crisis. In a recent report on water policy and protection in Africa, the UN Environment Programme highlighted the lack of political attention across Africa to its water crisis and the lack of leadership and policy on the subject. Congo-Brazzaville, Nigeria, and Sierra Leone don't even have a formal water policy; Cameroon says it has no one to champion the cause. Twenty-five countries, including Namibia, Swaziland, Rwanda, and Mozambique, said they did not have enough human capacity; Burundi said it had experienced too many changes of ministers to have a policy; and Ghana said it had trouble collecting revenue from local sources. Liberia said it had difficulty obtaining donor funds, and Libya and Zimbabwe said they did not have the infrastructure.[6]

At a May 2012 meeting in Rome, the United Nations Food and Agriculture Organization (FAO) adopted guidelines for safeguarding peoples' rights to own or access land, fisheries, and forests. The voluntary guidelines outline principles and practices that governments can use when making laws on land tenure. Said Alexander Müller of the FAO in a statement, "We know that the poorest people and the most vulnerable groups feel the first and the most pressure when it comes to issues like land-grabbing. Therefore, these voluntary guidelines protect human rights with a very clear focus on small-scale farmers, vulnerable groups, but also on women."[7]

While it is a positive step to see the UN recognizing the impact of land and water grabs on the most vulnerable, the voluntary nature of these guidelines is troublesome, says GRAIN. They say these guidelines, as well as ones being developed by pension funds, the private sector, and the World Bank, are seen as ways to regulate through voluntary codes and standards what is still considered an acceptable practice. The idea is to distinguish those land deals that meet certain criteria and can approvingly be called "investments" from those that don't. Voluntary self-regulation is ineffective, unreliable, and no remedy for the "fundamental wrongness" of these deals, says the group. Rather than help financial and corporate elites to "responsibly invest" in farmland, we need them to divest. Besides, says GRAIN, putting forward codes of conduct presupposes that the host countries of these land-purchase deals will adopt them. But in many cases governments use them to suppress the rights of their own people.[8]

In *Land, Life and Justice*, a scathing report on land and water grabs in Uganda, Friends of the Earth International says that poor governments such as the government of Uganda are desperate to attract foreign investment and don't want to look at the consequences. The group calls on African governments to conduct comprehensive research on the impacts of land grabs, protect natural forests rather than foreign-controlled tree plantations, create and enforce strict social and environmental policies on food production, domesticate international treaties and conventions regarding land and sacred sites, hold international financial institutions such as the World Bank accountable for funding projects that increase poverty through violation of community rights, and stop land and water grabs for agro-fuels, carbon-credit trading, and monoculture food systems, supporting agro-ecological farming instead.

Friends of the Earth calls for an international moratorium on more large-scale land acquisitions and for the return of plundered land. It says the international community can promote food and water security in Africa by helping to implement genuine agrarian and aquatic reform programs and targeting public investment to community and family farming. The group calls on governments to abide by their obligations under international human rights conventions, particularly the human right to food and water.[9]

Commons expert David Bollier points out that a major issue for small farmers and peasants in the Global South stems from different legal presumptions about what is considered to be property. Under European law, land must be registered and there must be formal title to it. But in rural Africa customary use rights in land are the norm. Conveniently for the investors, the European interpretation of property law is accepted in these land transactions, and this facilitates acquisition of legal title at cheaper prices.[10]

There are 1.5 billion small-scale farmers in the world who each live on less than two hectares of land, and many of them have no formal title to their land. It is estimated that 90 percent of the people in sub-Saharan Africa — some 500 million people — use the lands as a matter of custom and do not have statutory title to them. Bollier says this is the perfect place to implement laws and protocols that protect the commons, based on governance that can stand up to both private and state enclosures. Friends of the Earth International agrees, calling on governments to design legislation to protect citizens who own land under customary tenure options.

GROWING FOOD TO PROTECT WATER

Global food-production reform is essential if we are to save water for people and the planet. Local land and water, even in wealthier food-producing countries, must be used first and foremost to provide food for local and domestic markets and to maintain and restore healthy watersheds. Local and regional governments must prepare detailed inventories of their water supplies and set out long-term programs for conservation and restoration, including setting priorities for water access that may have to limit how much water is used for food exports.

Poor use of water, such as for biofuel production, must be phased out. Agriculture and water policy should be merged and food policy should promote water-retentive farming and water protection. Robert Sandford says that the key to balancing the trade-off between agricultural water use and availability for other needs requires transparent, well-coordinated, collaborative design and implementation of integrated water management and agriculture policies, at both the national and regional levels.[11]

Agriculture development goals and policies must be linked to water availability and sustainability, and at the moment few places in the world are doing this. Trade policy is set in one department, energy in another, agriculture in another, and water in yet another. So the energy and agriculture departments may promote heavily subsidized biofuel production, causing farmers to convert less water-intensive crops into water-guzzling biofuels, while the trade department is out looking for big new markets and writing deals to promote biofuel trade. Meanwhile, no one in any of these departments is talking to anyone who knows how much water is at stake and the impact of such policies on local watersheds.

David Schindler calls for "holistic watershed management" in order to protect water supplies generally and to stop the spread of eutrophication. It is not enough, he says, to improve nutrient removal; we have to better protect the catchments that supply the water. That will mean placing strict limits on the amounts of nutrients and chemicals that can be allowed in food production and the restoration of wetlands and riparian zones. This will require new ways of thinking about how we live and work, including comprehensive planning of human activity in the catchments that supply our water, which will in turn require governance on a watershed or catchment basis. We will also have to rethink livestock production and implement regulation of the disposal of livestock waste. Good water management can reduce water use in livestock production by 80 percent, but there are few rules to force farmers to use best water practices.[12]

Eutrophication is a true result of modern water. In mere decades the practice of growing food has so dramatically changed that 50 percent of people on the planet have come to depend on nitrogen fertilizers for their food. This poses a grave risk to lakes all over the world. If we do not grapple with this issue of nutrient runoff from industrial farming, says Schindler, "Our heads will be anointed with oil and algae when we go to the beach for a swim. Our cups will run over because no one will want to drink what is in them. And goodness and mercy will follow the Lord, but not us — for we shall dwell in the house of our wastes forever."[13]

We must also learn to adapt crop choices and production to better reflect local rainfall patterns and cut our dependency on irrigation where blue-water supplies are limited. We need strict laws to enforce state-of-the-art technology to reduce the water footprint of irrigated agriculture, and even to restrict or ban the use of irrigation in some areas and where it is used for food exports.

Sandra Postel says that getting more nutrition per drop can stretch domestic water supplies and reduce the need to seek land, water, and food from other countries. Drip irrigation, which delivers water directly to the roots of plants at very low volumes, can cut water use by up to 70 percent compared with old-style flood or furrow irrigation, she writes, while increasing crop yields by 20 to 90 percent. Although drip irrigation has expanded in some parts of the world in recent years, it is still used on only about 3 percent of irrigated land in China and India, the world's top two irrigators, and about 7 percent in the United States.

Monitoring and restricting access to groundwater is crucial. Pumping from the aquifers of the upper Ganges in India and Pakistan produces the world's biggest groundwater footprint by far, says Postel, followed by the aquifers of Saudi Arabia, Iran, western Mexico, the American High Plains, and the North China Plain. After the Texas legislature capped pumping from the Edwards Aquifer two decades ago, irrigation efficiency rose and the city of San Antonio and surrounding area cut its water use almost in half.[14]

The practice of irrigating drylands by using dwindling groundwater sources to produce intensive monoculture crops must be stopped. Industrialized intensive food production does not seek to manage existing water supplies wisely; rather, it is engaged in a never-ending search for new supplies, hurting both local water sources and land. Luc Gnacadja, executive secretary of the UN Convention to Combat Desertification, says there is a widespread but misguided belief that drylands are wastelands or marginal lands with low productivity and adaptive capacity, where poverty is inevitable. The perception is that they contribute little to national prosperity and yield no good returns on investment. Nothing could be further from the truth, he asserts. The fact is that drylands comprise one-third of the world's land mass and population, 44 percent of the global food-production system, and 50 percent of the world's livestock. In addition, dry forests are home to the planet's largest diversity of mammals, whose survival literally hangs on the arid-zone forests. Drylands hold the key to world hunger if treated properly.

If not handled with care, dryland suffers degradation and becomes acutely vulnerable to desertification, which does not allow even a blade of grass to grow. Like Schindler, Gnacadja calls for holistic planning in drought-plagued drylands, with a greater focus on the "forgotten billion" who, with support, hold the answer: small-scale community-based food farming. Restoring degraded land and returning to more traditional, biodiverse, and sustainable water-retentive agricultural practices will protect water and save lives.[15]

All these changes would dramatically reduce the global food market and force countries and communities to better care for their own water supplies. Pulling out of land and water grabs, combined with a necessary shrinking of the global food trade, will have positive impacts on richer countries, too, many of which are letting their best land be developed and urbanized as they import more food and virtual water. They will be forced to take better care of their existing water supplies and to renew a commitment to domestic land management — both key to a water-secure future.

CONFRONTING THE LORDS OF FOOD

However, none of this will be possible if we do not confront the global corporate chokehold on food production. A recent report by the ETC Group, a Canadian organization that monitors biotechnology worldwide, says that just six multinational gene giants control the current priorities and future direction of agricultural research worldwide. Syngenta, Bayer, BASF, Dow, Monsanto, and DuPont control 60 percent of commercial seeds, more than 75 percent of agrochemicals, and the lion's share of all private R&D in these sectors. They are teaming up with the world's two richest men, Bill Gates and Carlos Slim, to get bargain genetically engineered (GE) seeds and traits into the hands of farmers in the Global South, in the name of charity.

The notion that farmers in poor countries will benefit from GE seeds after their patents have expired is absurd, says Silvia Ribeiro, ETC's Latin American director. "Under the guise of charity, the Gene Giants are devising new schemes to soften opposition to transgenics and reach new markets. In reality, the Gene Giants don't have the capacity or the interest to supply the diversity needed in sustainable farming systems or to meet the urgent need for locally adapted varieties, especially in the face of climate change," she says, adding that these companies will ultimately control the terms of access even to expired seed patents, reinforcing their market power.[16]

The story repeats itself in other sectors. Five companies now dominate global grain trading, four of which — Bunge, Cargill, Continental, and Louis Dreyfus — also dominated a hundred years ago! Now, however, they are able to market their products through transnational supermarkets such as Walmart, and this means, says Greenpeace, that a few powerful companies dictate industry protocols to millions of small farmers, small suppliers, and consumers. Control of the food industry extends practically "from field to fork." Agribusiness claims to "feed the world," says Greenpeace, but instead engineers crops to depend on chemicals that those same companies sell. Today just twenty large agribusinesses control food in Mexico, making huge profits, while the country is now experiencing its worst food crisis in six decades.[17]

If we are to save the world's land and, therefore, its water by more local sustainable and natural farming, we must confront the "foodopoly" that Wenonah Hauter describes. This will require far-reaching legislation and regulatory changes that are part of a larger strategy for restoring truly participatory democracy. "Creating a just society where everyone can enjoy healthy food produced by thriving family farmers using organic practices can only be realized by making fundamental structural changes to society and to farm and food policy,"[18] she writes. These must include

- laws to break up the monopolies of the food and seed giants (as has been done in the past in the United States, for example) and curb their financial access to elected officials;

- reformed trade policies that would favour local food production and food security and restore watersheds;

- domestic and international regulations to rein in food speculation;

- an end to government subsidies to corporate farms and agribusiness;

- domestic policies that promote sustainable family farming and a fair return to the family farm gate;

- support for small and-medium sized independently owned and operated food enterprises to help invigorate rural economies;

- foreign aid policies that promote local communities and sustainable farmers in the Global South; and

- domestic and international food-quality standards based on the precautionary principle, to ensure that food safety is a top priority.[19]

As well, governments must promote management systems whereby the supply and demand of food are regulated and collectively marketed and imports are limited in areas where domestic products can meet demands. Canada has had supply-managed systems for wheat, barley, milk, eggs, and poultry for many decades, with great success for family farms and rural communities. The Canadian Organic Growers Association calls it "orderly marketing" and says it comes from the old-fashioned idea (old-fashioned in today's corporatized food world) that in the interests of both citizens and farmers, governments should attempt to stabilize and support agricultural economies. Marketing boards give family farmers collective power to negotiate prices with powerful multinational food processors and provide farmers with better prices than they would get if they were dealing directly with big corporate buyers.[20]

For decades the success of this system was a rare area of agreement among politicians of all stripes, as it nurtured a stable and successful industry in all of these sectors. The Canadian Wheat Board, one of the world's largest, longest-standing, and most successful state trading enterprises, kept giants such as Cargill largely out of the Canadian grain trade.

However, the Harper government recently dismantled the Wheat Board, against the wishes of the majority of its members, and has put the remaining supply-managed sectors up for grabs in upcoming trade agreements. This was a terrible mistake. Healthy farming in healthy rural economies not only provides the most stable food supply, it also protects water. If we do not challenge the current direction of food production in our world, we will see more devastated topsoil unable to produce food, more desertification, more starvation and its attendant human migration, and the continued loss of water.

SUPPORTING MINING RESISTANCE EVERYWHERE

There is another issue involving land and water that must be addressed: the metal mining industry, which is growing at a furious rate in the face of insatiable global demand. Currently mining is the second largest industrial user of water (not including agriculture) after power generation, reports industry journal *Global Water Intelligence*. The mining industry uses between 7 billion and 9 billion cubic metres of water annually, about as much water as a country such as Nigeria or Malaysia uses every year.[21] As well, each year, mining companies dump more than 180 million tons of hazardous waste into rivers, lakes, and oceans worldwide, 1.5 times the amount of municipal solid waste the United States sends to landfills every year.

In their 2012 report *Troubled Waters*, MiningWatch Canada and Washington-based Earthworks identify the companies that continue to use worst practices in their waste disposal, saying they are threatening vital bodies of water around the world with toxic chemicals and heavy metals. These tailings can contain as many as three dozen dangerous substances, including arsenic, lead, mercury, and cyanide. Many companies are guilty of a double standard, dumping their mine wastes into the rivers and oceans of other countries even when their home countries have bans or restrictions against the practice. Of the world's largest mining companies, only one has policies against dumping in rivers and oceans, and none has policies against dumping in lakes.[22]

Around the world, communities are resisting these mining operations, often facing threats, intimidation, jailing, beatings, torture, and even death from the companies or from local thugs operating on their behalf. Nowhere is the struggle more intense than in Latin America, blessed with an abundance of water and mineral wealth and the site of a mining invasion. *Bloomberg News Magazine* reports on a continent-wide conflict that pits South American governments and big foreign-based companies against local communities who stand to lose their homes and livelihoods as water is poisoned or diverted to industrial use. Many leaders across the region were elected on promises to fuel economic growth and lift their populations out of poverty; they are fast-tracking water-use approvals for mining, agribusiness, and other water intensive industry, says Michael Smith, the report's author.

As a result, Brazil's GDP increased 43 percent from 2002 to 2012, and the economy of Chile, where copper exports account for one-third of government revenue and where mining companies plan to spend another $100 billion by 2025, grew by 58 percent. Peru will expand by 6 percent in 2013, the fastest pace in South America, driven by investments in gold, silver, and copper mines. (In the last five years, over 200 people have been killed in mining clashes in Peru.)[23] More than three hundred corporations, many of them mining companies, have registered in Paraguay simply to access the waters of the Guaraní Aquifer, says Smith. He quotes Silvia Spinzi, the country's director of water resources, as saying that after registering, these firms just have to buy enough land to drill a well and remove water.[24]

The price of such development is very high. In March 2013 the drinking water that supplies 60 percent of Uruguayans was fouled by algae contamination in the Santa Lucia River, which is suffering from runoff from an open-pit iron mine and industrial agricultural pollution. Frequently small farmers, indigenous peoples, and villages are seeing their lakes and rivers dried up or poisoned, their livelihoods destroyed, and their communities abandoned because of foreign mining operations. Early in 2010, the Latin American Observatory of Environmental Conflicts reported that there were 118 mining conflicts in fifteen countries in Latin America, almost one-third of these involving Canadian companies. Canada is home to 75 percent of the world's mining companies, and an industry report found Canadian companies four times more likely to be at the centre of human rights and environmental conflicts than those from other Western countries.

In Mexico, where Canadian companies account for 204 of the 269 foreign mining companies, a series of assassinations of anti-mining activists has set the stakes very high for anyone resisting big mines and has made Canada a pariah in the international human rights community. In November 2009, activist and community leader Mariano Abarca Roblero, who had been in jail for protesting a barite mine run by the Canadian mining company Blackfire in his community of Chicomuselo, Chiapas, was shot dead. The assassins were all current or former employees of the company.[25] After much public outcry, the RCMP, Canada's national police force, launched an investigation into the murder.

In March 2012, Bernardo Vásquez Sánchez, of the Zapotec community of San José del Progreso, in Oaxaca province, was gunned down by assassins with links to the Vancouver company Fortuna Silver Mines. On October 22, 2012, lsmael Solorio Urrutia, leader of the local community activist group El Barzon in his community of Chihuahua, and his wife, Manuela Martha Solís Contreras, were shot and killed while driving in their truck. Solorio and his son Eric had been badly beaten months before, and other members of the group, set up to oppose the Cascabel Mine, a subsidiary of Vancouver-based MAG Silver, had been terrorized and threatened as well.

Blue Planet Project's Mexican organizer, Claudia Campero Arena, a human rights activist who works with the Coalition of Mexican Organizations for the Right to Water, laments that Mexico has become increasingly dangerous. Defending territory from transnational corporate interests is now a life-threatening activity. Grassroots dam, mining, and logging opponents and water defenders all face the risk of both criminalization by authorities and of becoming victims of threats, beatings, and murder. Asked where people get their courage to continue the struggle, Campero says that it comes with the "burst of indignation" that people experience when they understand the terrible injustice these projects represent for communities and for nature. "These very brave communities understand that these projects will change their future and their kids' and grandkids' possibilities for happiness, livelihoods and enjoyment of their territory."

If anything, the situation is worse in Guatemala. The oppression and intimidation have not stopped since the thirty-six-year civil war that ended in 1966. Successive Guatemalan governments have promoted aggressive development of the country's mining sector and looked the other way as human rights violations escalated in many of the 250 mining concessions opened in the past few years alone. Most are engaging in open-pit mining, mountaintop removal, and cyanide leaching processes to extract the gold and nickel from the ore, contaminating local waterways. In 2011 I visited several contentious Guatemalan mine sites as a guest of Grahame Russell of Rights Action, a Canadian who works tirelessly for justice in Guatemala, and met some of the victims of intimidation and terror.

I met members of a Q'eqchi' community who were terrorized for resisting the Estor nickel mine, until recently owned by the Canadian mining company HudBay Minerals. They told horrific stories of beatings, imprisonment, gang rapes, and murder committed by local thugs who provide security for the company. I was particularly touched by the story of German Chub Choc, a beautiful young man, father of a small boy and not active against the mine, who was playing soccer with friends one afternoon when a gang of security personnel employed by HudBay arrived to assassinate land-rights activist Adolfo Ich Chamán. They hacked and shot Chamán to death in broad daylight and then sprayed bullets indiscriminately, paralyzing German from the waist down.

We also went to the infamous Marlin Mine, run by Goldcorp of Canada. Its environmental and human rights abuses have been so well documented that the Inter-American Commission on Human Rights has called on the Guatemalan government to suspend operations at the mine. Again I met many victims who had stood up to the mine, but Diadora Hernández stands out for her exceptional courage. The company wanted her small property that abuts the mine, where she subsistence farms, but she refused to sell it. On July 7, 2010, a man hiding behind a tree on her property shot Diadora in the face, leaving her for dead. The local police refused to take her to hospital, so her daughter (accompanied by screaming granddaughter) had to take her in a taxi. Miraculously Diadora survived, but the water on her property has mysteriously dried up, forcing her to buy supplies from a private vendor.

There is no international law governing mining projects, says Shefa Siegel, a fellow in the politics of mining at the University of British Columbia, in a report called *The Missing Ethics of Mining*. Instead there are more than a dozen codes, all voluntary and all based on "corporate social responsibility," which is really an exercise in public relations between mining companies and local communities. Even though we are in the midst of a worldwide resource boom, there is very little discussion in official policy circles about how governments can set environmental and human rights standards for mineral extraction.

"Across these initiatives, the guiding principle is to promote economic development that benefits everyone involved — foreign companies, host governments, as well as local communities — not to question the underlying economic and ecological value of specific mines. The expansion of mining is accepted as inevitable," says Siegel.[26] John Briscoe, a Harvard professor and former advisor on water to the World Bank, says that governments are making the right decision in providing water to industries that benefit the majority of their populations, even if that means displacing some people. He says, "The value of the water in the mining industry is very, very high."[27]

There are now hundreds of organizations in Latin America and around the world working to stop the abuses of the mining industry. Some are directly involved in local resistance movements and others support them, raising public awareness and funds. More and more work is being done to get the story of these abuses out to the general public, especially in the home countries of the mining companies.

Groups are also promoting laws in the countries where these abusive mining companies are registered that would hold them more accountable to domestic standards. Canadian Member of Parliament Peter Julian has introduced a private member's bill that would improve access to Canadian courts for those who have suffered abuses from Canadian mining companies. With the support of Rights Action, a number of the victims of HudBay in Guatemala — including Angelica Choc, the widow of slain activist Adolfo Ich Chamán, and German Chub Choc — have instituted a $67-million lawsuit against HudBay in the Ontario Superior Court. They are represented pro bono by Murray Klippenstein, an extraordinary and committed Toronto-based justice lawyer, and his team.

Amnesty International supports the case, citing previous decisions by Canadian and British courts, as well as evolving international legal principles establishing that parent companies can be held liable for the actions of subsidiaries where the possibility of injury or harm is foreseeable. "Canadian society has a strong interest in ensuring that Canadian corporations respect human rights, wherever they may operate and whatever ownership and other business structure they may put in place to advance their operations," said Amnesty in its March 2013 submission to the court.[28]

Some countries are reining in the power of these foreign mining companies. Bolivian president Evo Morales ended five hundred years of foreign industrial domination by bringing in a new mining code clarifying that all minerals belonged to the people of Bolivia. Responding to strong anti-mine activism and violence at the Malku Khota Mine in Potosí, owned by Vancouver-based South American Silver, Morales nationalized the mine in 2012 and came to an agreement with the local indigenous community to proceed with its operation. Bolivia is sitting on the world's largest known lithium deposits, but now any new foreign investment must be partnered with the government, which will have the authority to impose environmental standards and insist that the mined lithium be used to create secondary manufacturing jobs in Bolivia.

In April 2013 the Appeals Court of Chile suspended work in the gold, silver, and copper Pascua-Lama Mine, owned by Canada's Barrick Gold, because the country's water department found that the operation was polluting groundwater. Located at an altitude of 4,000 metres in the Andes Mountains on the border between Chile and Argentina, Pascua-Lama sits at the headwaters of the Estrecho River, where it was causing "severe pollution," according to Lucio Cuenca, director of the Latin American Observatory of Environmental Conflicts. The lawsuit was filed by the local Diaguita indigenous communities, who said their local water supplies had been contaminated with arsenic, aluminum, copper, and sulphates.[29] On May 23, 2013, Chile's environmental regulator fined Barrick $16 million, the maximum allowable under Chilean law. A month later, Barrick shareholders launched a class action suit against the company, saying that it had made false statements and concealed material information on the Pascua-Lama operation.

In El Salvador, where a typical metal mine uses as much water in an hour as the average Salvadoran family takes twenty years to consume, the government has established a moratorium on metal mining to protect limited water supplies that local farming and fishing communities need to survive. Pacific Rim, another Canadian corporation, is taking the El Salvador government to court for delaying a permit for a gold mine that threatened to contaminate the largest river in the country and consume 30,000 litres of water a day, drawn from the same sources that currently provide local residents with water only once a week. The company is suing El Salvador for US$315 million in compensation and the dispute is before the International Centre for the Settlement of Investment Disputes at the World Bank. The government of El Salvador, however, is remaining strong in its belief that it has the right to set its own environmental and mining polices, and it is backed by the international mining- and water-justice movements.

In its 2013 report *Mining for Profits in International Tribunals*, the Washington-based Institute for Policy Studies found that corporate lawsuits against foreign governments relating to oil, gas, and mining are on the increase, and that Latin America is being particularly targeted.[30] At a June 2012 rally outside Pacific Rim's headquarters in Vancouver, the institute's director, John Cavanagh, said, "The world should applaud the efforts of the Salvadoran people to safeguard their country's long-term health and prosperity by becoming the first in the world to ban gold mining. Instead, this ruling is one more example of international investment rules undermining democracy in the interest of short-term profits for foreign investors."

While these stories show that some progress has been made, much still needs to be done. Lynda Collins, who teaches human and environmental rights at the University of Ottawa, says that one tool we can use is the newly recognized human right to water. Collins says that states now have an international obligation to refrain from causing violations of the right to water outside their own territories, and this extends to a responsibility to regulate the extraterritorial conduct of their corporate nationals. A country must refuse to fund activities of its corporations that are likely to violate the right to water in another country, and must refrain from interfering with the ability of that country to effectively regulate the environmental conduct of multinational corporations within its borders. In particular, the corporation's home state should refrain from concluding investment treaties that penalize host states for measures designed to protect public health and the environment, says Collins. Further, a home state may have a positive obligation to regulate the conduct of its corporate nationals extraterritorially.[31]

That the UN resolution affirming the rights to water and sanitation could have such far-reaching impacts is exciting. The right to water means that a transnational mining corporation does not have the right to harm or destroy local water sources upon which people depend. We must build upon this to curtail the criminal activities of some mining operations across the globe. This is a foundation to act upon.

16

A ROAD MAP TO CONFLICT OR TO PEACE?

If there is a water war, it will not be the water that caused war, but rather a war that was in search of an issue and found water.
— Ami Isseroff, late Israeli peace activist[1]

THE CENTRAL ISSUE OF our time is whether we will see the dwindling resources of our planet as a cause for competition, leading inevitably to conflict, or whether we will see them as a means of co-operation, leading to peace.

A 2012 U.S. intelligence report warned that water shortages, polluted water, and floods will increase the risk of instability in many nations important to the country's national security interests. This could lead to state failure and increased regional tensions, and even be a tool for terrorists. The report zeroes in on seven key river basins located in the Middle East, Asia, and Africa — the Indus, Jordan, Mekong, Nile, Tigris-Euphrates, Amu Darya, and Brahmaputra basins — and warns that if conflicts in these areas are not resolved, water could be used as a weapon by more powerful upstream neighbours' impeding or cutting flows. The report, prepared at the request of the State Department, recognized that in the past, water problems have more often than not led to water-sharing agreements rather than to violent conflicts. But the authors, drawing on reports from a variety of government intelligence agencies, warn that as water shortages become more acute, this could change.[2]

Canada's Department of National Defence agrees with this assessment. It says that up to sixty countries could fall into the category of "water scarcity or stress" by 2050, making water a key source of power and a basis of future conflict. The danger of "resource wars," between states and within them, is acute, reports the department, which predicts violence in the developing world in particular.[3] Between 1980 and 2005 in sub-Saharan Africa, there were twenty-one civil conflicts, sixteen of which involved water and five in which water was the sole issue.[4]

GROWING WATER TENSIONS IN ASIA

An alarm was sounded at a May 21, 2013 summit of Asia Pacific heads of state in Bangkok. One leader after another rose to warn that fierce competition for water could trigger conflict in the region unless nations co-operate to share the diminishing supply. From Central to Southeast Asia, regional efforts to secure water have sparked tensions between neighbours that rely on rivers to sustain booming populations, reports Agence France-Presse. Brunei's Sultan Hassanal Bolkiah said that competition for water could lead to international disputes. Urbanization, climate change, and surging demand for water from agriculture have heaped pressure on scarce water supplies while the majority of people in Asia Pacific still lack access to safe water, despite strong economic growth.[5]

The 4,660-kilometre Mekong River is already one source of international tension. As the water level in the Mekong dips to historic lows, concerns are rising over Chinese dams operating or under construction upriver from communities in Burma, Cambodia, Laos, Thailand, and Vietnam that are experiencing dropping water levels. "China is fast failing the good-neighbour test," says an editorial in the *Bangkok Post*. "The trouble is China's unilateral decision to harness the Mekong with eight hydro-electric dams."[6] Energy-starved China is using electricity generated by the dams to fuel its rapid economic growth, without regard to the adverse impact on its neighbours, declared Vietnamese president Truong Tan Sang at an August 2012 APEC summit. The World Wildlife Fund notes that, for the first time in several thousand years, the Mekong Delta in Vietnam, where 18 million people live, is shrinking.[7]

Remapping of the water flows in the world's most heavily populated and thirstiest region is happening on a gigantic scale, with potentially strategic implications, says the Associated Press. On the eight great Tibetan rivers alone, almost twenty dams have been built or are under construction, while another forty are planned. While China is not the only culprit, it is blamed for everything from sudden floods to water-depleted rivers and lakes in vulnerable towns and capitals from Pakistan to Vietnam. The fear is that China's accelerating program of damming every major river flowing from the Tibetan plateau will trigger natural disasters, degrade fragile ecologies, and divert vital water supplies.

The Brahmaputra River begins in southwestern Tibet (where it is known as the Yarlung Tsangpo) and flows for 1,600 kilometres south until it makes a sudden U-turn just before it enters India. It is from this "grand bend" in the river that China has talked about diverting water as part of its north–south water diversion scheme, which involves three manmade rivers carrying water from the Tibetan plateau to China's arid north. "Whether China intends to use water as a political weapon or not, it is acquiring the capability to turn off the tap if it wants to — a leverage it can use to keep any riparian neighbours on good behaviour," says Brahma Chellaney, an analyst at New Delhi's Centre for Policy Research and author of the book *Water: Asia's New Battlefield*. He believes the issue is not whether China will reroute the Brahmaputra River, but when — tantamount to a declaration of war on India.[8]

The ex-Soviet states of Central Asia are engaged in an increasingly bitter standoff over water resources, reports Agence France-Presse journalist Akbar Borisov, adding more instability to the volatile region neighbouring Afghanistan. Plans in Tajikistan and Kyrgyzstan for two of the world's biggest hydroelectric power stations have enraged their downstream neighbour Uzbekistan, which fears losing valuable water. Russia is being pulled into the dispute, which dates back to the allocation of resources when the Soviet Union broke up in 1991. Borisov quotes Uzbek president Islam Karimov, who said on a September 2012 visit to Kazakhstan that this battle over water resources could sharpen tension in the region to such an extent that it could spark not just "serious resistance but war."[9]

WATER AS WEAPON IN THE MIDDLE EAST

Water has been a source of conflict and used as a weapon of war for decades in the Middle East. The ancient Mesopotamian Marshes were drained during the Iran–Iraq war of the 1980s to give the Iraqis a tactical advantage. Saddam Hussein drained them further during the 1990s in retribution against Shias who hid there and against the Marsh Arabs (Ma'dan) who protected them. By the early 2000s the marshes were just 10 percent of their original size. The UN says the number of Ma'dan, who had a population of about half a million in the 1950s, has shrunk to about 20,000, and as many as 120,000 have fled to refugee camps in Iran.

Still, by the mid-2000s the marshes had been greatly rejuvenated, and many of the Ma´dan were returning, thanks largely to people such as Peter Nichols of Waterkeeper Alliance; Azzam Alwash (who grew up in the marshes) of Nature Iraq, an environmental organization devoted to restoring the country's waterways; and Michelle Stevens of Hima Mesopotamia, an international NGO dedicated to restoration of the marshes. But plans to build a chain of twenty-three dams upstream along the Turkey–Syria border are once again putting the region under threat. Already water diversions are bringing back drought, salinity, and hunger.

While Turkey claims that the massive dam expansion is for hydroelectric power, the waters of the Mesopotamian Marshes are again being used as a weapon. American journalist Jay Cassano says that the real reason for the dams is to flood the canyons where the Kurdistan Workers' Party mobilizes near the mountainous Iraq–Turkey border. The government has made no secret that the strategically placed dams will form a massive wall of water close to Turkey's border with Iraq, making the terrain impossible to pass by foot. As well, by virtue of being upstream from Iraq and Syria on both the Tigris and Euphrates Rivers, Turkey effectively controls the flow of water southward.[10]

The privatization of Egypt's water and its diversion to the wealthy was a key factor in the "Arab Spring" uprising against the Mubarak government, says English professor Karen Piper of the University of Missouri. Within months of privatization in 2004, the price of water doubled; thousands who could not pay their bills had to go to the city outskirts to collect water from the dirty Nile River canals. In 2007, protesters in the Nile Delta blocked the main coastal road after the regional water company diverted water from farming and fishing towns to affluent resort communities. The demonstrations grew in intensity and melded into the larger liberation movement. Egyptian water activist Abdel Mawla Ismail said, "Thirst protests...started to represent a new path for a social movement." From this path, says Piper, the revolution that consumed the nation in 2011 seemed inevitable.[11]

"Criminal mismanagement" of the country's water by the al-Assad regime in Syria was also one of the root causes of the current uprising in that country. When Bashar al-Assad took over in 2000, he opened up the regulated agriculture sector for big farmers, many of them government cronies, to buy up land and drill as much water as they wanted. This severely diminished the water table and drove small farmers and herders off the land, Syrian economist Samir Aita told American journalist Thomas Friedman.[12] The Washington-based Center for Climate and Security reports that from 2006 to 2011, 60 percent of the country's land experienced "the worst long-term drought and most severe set of crop failures since agricultural civilizations began in the Fertile Crescent many millennia ago."

The drought and food crisis that followed displaced close to a million people and drove close to 3 million into extreme poverty. The Assad regime also denied licences to allow Kurdish communities in the northeast to draw modest amounts of water from wells; this exacerbated the exodus of hundreds of thousands to urban slums in the cities of the south, such as Aleppo, that became the centre of the first protests.[13]

"Young people and farmers starved for jobs — and land starved for water — were a prescription for revolution," writes Thomas Friedman. Adds the *Bulletin of the Atomic Scientists*, "The drought in Syria is one of the first modern events in which a climatic anomaly resulted in mass migration and contributed to state instability. This is a lesson and a warning for the greater catalyst that climate change will become in a region already under the strains of cultural polarity, political repression, and economic inequity."[14] Tragically, the huge influx of Syrian refugees into Jordan is straining that country's limited water supplies, leading to growing hostility against the newcomers.

Middle East watchers warn that the massive irrigation project known as the "Great Man-Made River," built by late Libyan dictator Muammar Gaddafi to tap into groundwater reserves in the desert, could also become a source of conflict. Consisting of an astonishing 5,000 kilometres of pipelines from more that 1,300 wells drilled up to 500 metres deep into the Sahara, the $30-million project could turn Libya into a major agricultural producer. But the waters of the Nubian Sandstone Aquifer System (NSAS) lie under the countries of Chad, Egypt, and Sudan as well as Libya, and the fossil waters it contains are non-renewable. "In a nutshell," says News Central Asia editor Tariq Saeedi, "whoever controls NSAS controls the economies, foreign policies and destinies of several countries in the region, not just north-eastern Africa." Adds Middle East–based journalist Iason Athanasiadis, "In a desertifying region already wracked by water conflict, Libya's enormous aquatic reserves will be a large prize for whoever gets the upper hand in this struggle."[15]

Gaza may not be "liveable" by 2020 and its aquifer may be unusable as soon as 2016, says an August 2012 report from the UN children's fund. The report highlights Gaza's acute water crisis, noting that the aquifer may suffer irreversible damage by 2020.[16] More than four decades of Israeli occupation have made it impossible to develop or maintain infrastructure for water. With no replacement parts available, broken pipes allow raw sewage to leak into groundwater. Salt and toxic nitrates from the Mediterranean also contaminate the water supply, already over-pumped for Israeli settlements in Gaza. Many houses have running water only once a week, and some families have no taps in or near their homes. In Gaza there is now no uncontaminated water, reports Victoria Brittain in the *Guardian*. Half of newborn babies are at immediate risk of nitrate poisoning, called "blue baby syndrome."[17]

UN special rapporteur Catarina de Albuquerque says, "This reality is a grave threat to the health and dignity of the people living in

Gaza and immediate measures are required to ensure full enjoyment of the rights to water and sanitation. Israel must facilitate the entry of necessary materials to rebuild the water and sanitation systems in Gaza, as a matter of priority."[18] The crisis is not just an ongoing violation of the human right to water of Palestinians living in Gaza, it also threatens the water supplies of the region, as the aquifer supplies water to Israel and Egypt as well.

SHARING WATERSHEDS

Can these and other conflicts be turned around through the need to share for survival? Yes, says a group of scholars who have examined nation-state conflicts involving water. Aaron Wolf, Annika Kramer, Alexander Carius, and Geoffrey Dabelko report that international water disputes — even among fierce enemies — have generally been resolved peacefully because water is so important that nations cannot fight over it. In fact, they say water has fuelled greater interdependence and helped build trust between warring parties.

Researchers at Oregon State University compiled data on every reported water-driven interaction between two or more nations in the past half-century and found that the rate of co-operation overwhelmed the incidence of acute conflict: 1,228 to 507. Wolf, Kramer, Carius, and Dabelko warn that talk of "water wars" will allow the military and other security groups to take over negotiations and push out development partners, aid agencies, and the UN. Water management offers an avenue to peaceful dialogue between nations, says the group, even when combatants are fighting over other issues, and water co-operation forges people-to-people connections.[19]

There are 276 river basins that are shared by two or more countries and about 300 agreements between states around shared rivers. While the Pacific Institute's Peter Gleick agrees that most water cross-border disputes are resolved diplomatically, he points out that climate change and the growing demand for water will increase the risk of conflict over international shared freshwater resources. "Climate change will inevitably affect water resources around the world, altering water availability, quality, and the management of infrastructure," he says. "New disputes are already arising in trans-boundary watersheds and are likely to become more common. The existing agreements and international principles for sharing water will not adequately handle the strain of future pressures, particularly those caused by climate change."

Gleick calls for the creation of cross-border water agreements where they do not now exist — more than half the world's international watercourses are not covered by co-operative management frameworks — and expansion of the scope of existing agreements to include all the elements of the hydrologic cycle, including aquifers. Joint trans-boundary research on the impact of climate change on shared water systems, as well as shared monitoring programs, can also lead to co-operation when tensions rise.[20]

In the case of shared international watercourses, states have obligations grounded in the international law governing contamination and depletion of transboundary water resources. Further, says the University of Ottawa's Lynda Collins, the right to water itself may impose an independent obligation on the part of a co-riparian state to preserve the quantity and quality of shared watercourses.[21]

It is crucial that nations ratify the UN Convention on the Law of Non-navigational Uses of International Watercourses, adopted in 1997 to help conserve and manage water supplies for future generations and resolve conflicts over shared watercourses. The convention would require states to prevent and reduce pollution in shared watercourses, establish a level playing field among watercourse states, incorporate social and environmental considerations into the management of the watercourse, and seek peaceful settlement of disputes.

As of 2013 the convention has received only thirty ratifications, five short of the number needed to bring it into force, and far short of the number needed to make it truly effective. In 2006 the World Wildlife Fund launched a global campaign to promote the convention and accelerate the ratification process. The Fund agrees with Gleick that most of the world's transboundary water resources still lack sufficient legal protection, without which it will be difficult for watercourse states to cope with future threats from human pressure and environmental change.[22]

A similar treaty to protect shared groundwater is also urgently needed. UNESCO's International Groundwater Resources Assessment Centre reports that there are at least as many, and likely more, aquifers that cross nation-state boundaries as rivers. Dr. David Brooks, of Canada's International Institute for Sustainable Development, says that rather than a simple quantitative sharing of water resources, true equity is achieved by taking into account the demands of everyone's needs rather than their traditional claim to the water. Brooks is clear that, given the sensitivity of aquifers to pollution and the near impossibility of decontamination, it is crucial to assert the principle of "no significant harm," that is, that each nation must agree not to pollute the shared waters.[23]

In 2008 the UN International Law Commission and UNESCO presented a draft Convention on Transboundary Aquifers to the UN General Assembly. The convention would require that aquifer states not harm existing aquifers and co-operate to prevent and control their pollution. It needs to be strengthened to address recharge and the need for restrictions on taking more water than a groundwater source can replenish. The convention has been stalled inside the United Nations and must be energized. Aquifers are supplying more of our water needs every day; they desperately need the protection of a binding international treaty.

CREATING PEACE THROUGH WATER

Peace-building is based on the practice of identifying the conditions that can lead to a sustainable and workable peace between adversaries, and it has a long history in the resolution of conflict between nations. Environmental peace-building seeks to resolve conflict through concern for shared ecosystems. In some cases it is used to stop the environmental degradation that is a by-product of conflict. War and violence often devastate the environments of all parties concerned, so there is great motivation to stop the conflict to restore food and water supplies. In other cases, fear over declining groundwater levels in a shared aquifer or the pollution and deterioration of a shared river can bring together sides that are in conflict over other issues. If the conflict itself is over water, it is resolved through common resource management. If the conflict is about other issues, water can become a tool to build bridges while the longer peace process takes place.

Friends of the Earth Middle East was founded in 1994 as a meeting place for Palestinian, Egyptian, Jordanian, and Israeli environmental NGOs to work together to protect and restore the Jordan River Valley, the Dead Sea, and the Gulf of Aqaba. The group promotes protection rather than development of the Dead Sea, seeks its designation as a UNESCO site, and opposes the proposed mega-project to channel water from the Red Sea to the Dead Sea. It has launched the Good Water Neighbors project to partner communities from warring factions over water education and co-operation.

In May 2013 the Israeli Water Authority removed a forty-year blockage of the Jordan River and started releasing water from the Sea of Galilee to replenish the Lower Jordan. Friends of the Earth Middle East, and others who have lobbied for this development, are cleaning the polluted riverbed and treating water to remove decades of pollution. This group says that environmental peacemaking builds trust, acts as a lifeline during conflict, and creates a shared regional interest in building an ecological community.[24]

The work of Friends of the Earth Middle East and other organizations has yet to translate into any kind of resolution of the water and human rights crisis in Gaza. To attempt to break the impasse, David Brooks, then of Friends of the Earth Canada, and Julie Trottier, of the Université Paul Valéry in France, created a plan to be used as background for the Geneva Initiative, a non-governmental effort to promote peace in the Israel–Gaza conflict. Noting that water governance is less a technical issue than a political one, they challenge the rigidity of the traditional quantitative approach to sharing water. Water is not like land; it can neither be described nor divided as if it were a pie.

Brooks and Trottier say that when a quantity of water "must" be received according to a treaty, it comes to be seen as an issue of national security, and any talk of changing the formula is seen as a threat to a state's authority. Their plan seeks a more appropriate concept of water rights so that demand is matched to the reality of existing supply and both parties recognize their interdependence in sustaining the quantity and quality of all shared waters. The authors recommend a set of principles that include equality and respect for the contrasting approaches with which Israelis and Palestinians have tackled water management.[25]

BUILDING INTERDEPENDENCE TO SHARE WATER

Seeing joint watercourses in this way challenges us to think beyond political boundaries and nation-state interests. Sandra Postel reminds us that rivers pay no mind to political boundaries, and if we want to sustain ecosystems in the face of growing food and energy demands, we will need to think more like watersheds and less like states or nations. We can optimize all the benefits that rivers give us only if we work together across borders to secure and share them. This takes, she says, a quantum leap in co-operation.

For example, a recent agreement between Mexico and the United States allows the U.S. to retain more than its share of the Colorado River in order to return flows to the parched Colorado Delta. Mexico will take less water during times of drought and be allowed to store water in Lake Mead, a U.S. reservoir, during times of surplus if it cannot use its entire allocation. In exchange, the U.S. will fund repairs of the damage that Mexico's irrigation canals suffered in a recent earthquake. "We share a culture, a border and resources with our neighbors in Mexico," said Michael Connor, commissioner of the U.S. Bureau of Reclamation, "and it's entirely appropriate that we also share solutions to the challenges we face in the Colorado River basin."[26]

A similar sharing arrangement could provide a peaceful solution to the contested Nile. A 1959 treaty between Egypt and Sudan divided up the flow of the Nile but did not allot any water to the other basin countries. What if instead, Postel asks, the Nile Basin nations traded the benefits produced with Nile water, such as electricity, food, fisheries, wetlands, and wildlife habitats? Store Nile water in the Ethiopian highlands, where evaporation rates are about a third as high as at Egypt's Lake Nasser, and the saved water could benefit all nations in the watershed. Regional investment in efficient water use, small-scale irrigation, and food sharing could provide Ethiopia with an alternative to opening its territory to land and water grabs.[27]

Unfortunately, in June 2013 Egypt's former president Mohammed Morsi said that Ethiopia's construction of a $4.2-billion hydroelectric dam on the Nile threatens his country's water security, and if Egypt loses one drop of water as a result of the dam, "our blood is the alternative."[28] While Ethiopia undoubtedly feels frustrated at being left out of the original treaty, it is building the dam to provide water for the industrialized farms and plantations that have replaced local communities through land grabs. Survival International says the dam and land theft threaten the lives and livelihoods of eight different tribes along the Lower Omo River.

There is an emerging movement to promote watershed governance, a concept based on the notion that in order to effectively protect water, it is necessary to do so on a watershed-wide basis. Rivers, lakes, and aquifers often cross political boundaries and need co-ordinated and co-operative governance among all the political jurisdictions involved. The POLIS Water Sustainability Project at the University of Victoria, British Columbia, promotes watershed governance in Canada to help governments and communities protect their shared watercourses. POLIS advocates for integrated long-term watershed-based planning and ecosystem-based legal reform as the best approach to water management and decision making. Watershed governance is based on the principles of conservation, stewardship, sustainability, and co-operation. Clearly this is easier to attempt within national borders than across them; asking countries to give up some sovereignty is difficult in today's world of regional tensions and rivalries.

The main problem in the Central Asian conflict, says the International Fund for Saving the Aral Sea, is the national policy interests of the region's nation-states. They see the issue of resolving problems only through the prisms of energy and water self-sufficiency and sovereignty, and that makes true co-operation impossible. The fund urges regional and ecosystem co-operation based on shared history and culture. Each country's interests and concerns must be heard by the whole, and compromises must be made. Not only will effective co-operation around shared watercourses save those waterways, it can resolve conflict on other issues. As a matter of survival, regional unity must become a national priority of all Central Asian states.[29]

Similarly, the Mekong River Commission, formed by an agreement among Vietnam, Cambodia, Lao PDR, and Thailand (China and Myanmar attend as observers) to jointly manage the river, has been widely criticized for having little real power to stop the destruction of the basin. Oxfam Australia is part of a global campaign group called Save the Mekong that works to ensure that communities know their rights regarding development decisions that affect their environment and their access to the Mekong's resources. In a report on the commission, Oxfam notes that it is not a supranational body and has no regulatory or enforcement power; it is really all about what economic benefit each country can "extract" from the Mekong. "At present, the MRC is a relatively small player sandwiched between larger ones in the business of promoting water related sustainable economic growth and development in the Mekong region,"[30] it says.

As a result, the commission has not been able as a unified body to open a serious dialogue with China over its dams. Edward Grumbine, visiting American senior scientist at the Chinese Academy of Sciences, warns that diplomacy is needed in the Mekong dispute and that countries will have to give up national interests for regional ones. "Given the enormous demand for water in China, India and Southeast Asia, if you maintain the attitude of sovereign state, we are lost," he says. "Scarcity in a zero sum situation can lead to conflict but it can also goad countries into more cooperative behaviour."[31]

MAKING WATERSHED GOVERNANCE WORK

In 1992 the United Nations Economic Commission for Europe ratified the Convention on the Protection and Use of Transboundary Watercourses and International Lakes to strengthen national measures for ecologically sound management of the continent's shared waters. Water sources in Europe cross national borders. More than

150 major rivers and 50 large lakes run along or straddle the border between two or more countries. Twenty European countries depend for more than 10 percent of their water on neighbouring countries, and five draw 75 percent of their water from upstream neighbours. The commission said that the convention takes a holistic approach based on the understanding that water plays an integral role in protecting ecosystems, and it represents an integrated water resources–management narrative that replaces an earlier focus on localized sources of pollution. The convention obliges member nations to use their shared waters in a reasonable and equitable way and to ensure their sustainable management.

Almost two decades later, the continent took a further step towards shared water governance. Europe's 2000 Water Framework Directive is a landmark policy initiative that establishes new requirements for integrated river-basin management in Europe and commits its member states to achieve good qualitative and quantitative status of all water bodies by 2015. The directive set up river basin districts: areas of governance designated according to river-basin boundaries rather than political ones. River basins are managed according to a plan that includes targets and timelines for restoration. The process is meant to be transparent and accountable to the public.

Cross-border collaboration has already been proven to work. The once badly polluted Lake Constance, the second largest alpine lake in Europe, whose catchment area lies in Austria, Germany, Liechtenstein, and Switzerland, has been restored as a result of co-operation among the countries and communities around the lake. However, says the European Environmental Bureau, a federation of more than 140 environmental citizens' organizations, lack of participation from a number of countries, lax enforcement of regulations, and lack of the promised public transparency all undermine the original goals of the directive.[32] The goals of sustainability and ecosystem protection have come up against the power of the private sector in

Europe and the inability of governments to protect the environment in a time of enforced austerity measures.

A network of citizen, community, and First Nations groups in Canada and the United States has come together to have the Great Lakes declared a commons, a public trust, and a protected bioregion. They are calling on governments to establish a watershed governance framework to protect the lakes in the face of grave threats. Pollution from industrial farming and industry, invasive species, climate change, and over-extraction threaten five bodies of water that hold more than 20 percent of the world's surface fresh water. The Great Lakes are (badly) managed by two nations, Canada and the United States, eight states, two provinces, and hundreds of municipalities. They cry out for common watershed standards, regulation, and enforcement.

The groups concerned posed a question: "What if the people who live around the Great Lakes decided to protect them based on the principles that informed the First Peoples in the region; namely, that the lakes must be shared equitably and fiercely protected for seven generations to come?" We decided the answer would be that the Great Lakes Basin must be seen as one watershed, and that all activity, public and private, must operate within a mandate whose goals are restoration and preservation of the waters of the basin. Wisconsin-based environmental attorney Jim Olson, of the group FLOW for Water, points out that all the waters of the Great Lakes are subject to public trust law, but in practice far too many governments and industries see the Great Lakes either as immune to depletion or as a commodity to be exploited.

Alexa Bradley, of the group On the Commons, says that for some the Great Lakes represent an open resource. This attitude allows privatization, appropriation, and entitlement to use and misuse the water, and the prioritization of market economics over ecological and justice considerations. "By its nature this resource grab is anti-democratic and undercuts both environmental protection and the equitable sharing of water. This exploitation makes the case for not just better water policy, but for a different kind of governance."[33]

Jim Olson argues that the public trust doctrine could be integrated into the Boundary Waters Treaty between Canada and the United States as well as the Great Lakes Water Quality Agreement to integrate water quantity and quality and ecosystem protection. Entrenching the waters as a public trust in these agreements would provide a basis for the International Joint Commission, which oversees joint management of the lakes, to demand accountability from all governments involved. Public trust principles would assert the inalienable right of public use as a safeguard against unforeseen claims by private interests such as diversion or exports. Using public trust arguments, groups around the lakes are successfully challenging fracking operations.[34] Emma Lui, water campaigner for the Council of Canadians, points out that Quebec has a moratorium on fracking and that more than a hundred New York municipalities have banned gas drilling.

The network wants a full treaty among all relevant governments to set consistent, basin-wide laws, regulations, and definitions to protect a new "Great Lakes Basin Commons." The process is beginning in the cities, towns, villages, hamlets, and farms that ring the Great Lakes and with the people who live on and love them, and there is growing grassroots support. Waterkeeper Alliance, an international network of local water-keeper organizations founded by Robert Kennedy Jr., has endorsed the model, both for the Great Lakes and for other watersheds.

At a November 2011 conference on the future of water in Mercosur, the southern common market, I urged the four countries that share the Guaraní Aquifer — Brazil, Paraguay, Uruguay, and Argentina — to come together and promote a similar model to protect it from abuse. In 2010 the four countries had signed one of the first aquifer agreements in the world, committing themselves to a dispute resolution process and making modest promises to curb pollution. But the agreement replicates similar transboundary surface-water treaties that protect the rights of the individual states and do little to oversee the aquifer as a shared watercourse or a public trust.

I called for a Guaraní Watershed Covenant among the nations to define the aquifer waters as a common heritage, a public trust, and a human right and to designate the waters as a bioregion to be strictly protected under a common set of laws. "In case I have not been clear," I said to a very receptive audience of policymakers, government officials, academics, and activists, "you are sitting atop a vast reserve of water in a very thirsty world, a reserve that is vital not only to the health and future of this region but to all of humanity. It is a treasure that must be protected by the four governments on behalf of the people and ecosystems of this region.... You must urgently take steps to protect the jewel that is the Guaraní Aquifer from plunder, for if you do not, plunder is coming."

TRUSTING WATER TO THE PEOPLE

All water is local. Communities that live on a watershed know it best, and their knowledge is irreplaceable. A crucial principle of watershed governance is that water is a public trust, and those who are affected must have a meaningful way of participating in decision making. Water policy is too often made at the top, without consultation. Sharing decision-making responsibility with local watershed communities will improve the way water is governed. Save the Mekong says that failure to consult with the public in any meaningful way has been central to the problems plaguing the Mekong River Commission. Hundreds of local cross-border groups are working to find real and long-term solutions that hold at least part of the answer to a water-secure Mekong Basin region.

Indigenous peoples have much to teach about water's care and preservation. In Latin America, indigenous resistance to water privatization, big dams, and destructive mines has been crucial in forcing governments to start dealing with water management and human rights in a different way. American Native organizations were a key component of the process that guided the Mexican–U.S. agreement to restore the Colorado River. First Nations opposition to the Northern Gateway Pipeline, intended to ship Alberta tar-sands bitumen to ports in British Columbia, has likely sealed its fate. In all these and countless other cases, indigenous peoples cite not only their direct dependence on local water systems for survival but also their spiritual and historic connection to these waters as the reason for the strength of their commitment.

The quest to protect water forever is also inextricably bound with human rights. If we want to transform conflict into peace and create new ways to govern that honour watersheds and ecosystems, it is essential to recognize that lack of access to clean water is a form of violence and that there can be no peace or good governance without justice. This means putting the human rights to water and sanitation, as well as the right to engage in the process, at the centre of a new, more collaborative form of watershed governance. Conflict transformation goes beyond the concept of conflict resolution: it requires confronting the unjust social structures that underlie the conflict.

In continuing to build and expand our international movement to protect water for people and the planet, we need to recognize how very far we have come in the past few years in the struggle for water justice. While much remains to be done, we have moved this issue forward in ways we dared not dream of just a few years ago. The global water crisis can be solved if we move steadily forward to protect water as a public trust and ensure just and equitable access. Water can be nature's gift to humanity in ways we have yet to understand.

NOTES

PRINCIPLE ONE: WATER IS A HUMAN RIGHT

1. THE CASE FOR THE RIGHT TO WATER

1. Oscar Olivera, in *The Corporation*, film by Joel Bakan and Mark Achbar, 2003.

2. Amnesty International, "United Nations: Historic Re-affirmation That Rights to Water and Sanitation Are Legally Binding," IOR 40/018/2010 (October 1, 2010), http://www.amnesty.org/en/library/asset/IOR40/018/2010/en/34b48900-9ee4-4659-8a63-9e887112e5f7/ior400182010en.html.

3. Oxfam International, "The Cost of Inequality: How Wealth and Income Extremes Hurt Us All," 02/2012 (January 18, 2013), http://www.oxfam.org/sites/www.oxfam.org/files/cost-of-inequality-oxfam-mb180113.pdf.

4. Statement released by UN Secretary General Ban Ki-moon, based on reports for several UN agencies, New York, World Water Day, March 22, 2010.

5. Shekhar Kapur, "Whose Water Is It Anyway?" *Tehelka.com* 10, no. 16 (April 20, 2013), http://tehelka.com/whose-water-is-it-anyway/.

6. Marc Bierkens, International Groundwater Resource Assessment Centre, "Groundwater Depletion Rate Accelerating Worldwide," September 23, 2010, http://phys.org/news204470960.html.

7. United Nations Environment Programme, *Towards a Green Economy: Pathways to Sustainable Development and Poverty Eradication* (Stockholm: United Nations, 2011).

8. David Blair, "UN Predicts Huge Migration to Rich Countries," *Telegraph*, March 15, 2007.

9.   Peter H. Gleick, *The World's Water, 2008–2009: The Biennial Report on Freshwater Resources* (Oakland, CA: Pacific Institute, 2009).

10.   Colin Chartres and Samyuktha Varma, *Out of Water: From Abundance to Scarcity and How to Solve the World's Water Problems* (Upper Saddle River, NJ: Pearson Education, 2011).

11.   Arjen Y. Hoekstra and Mesfin M. Mekonnen, "The Water Footprint of Humanity," *Proceedings of the National Academy of Sciences of the United States of America* 109, no. 9 (February 2012): 3232–37.

12.   George Monbiot, "Population Growth Is a Threat. But It Pales Against the Greed of the Rich," *Guardian,* January 29, 2008.

13.   Charles Vörösmarty, Peter McIntyre, et al., "Global Threats to Human Water Security and River Biodiversity," *Nature* 467, no. 334 (November 11, 2010): 555–61.

14.   Lester R. Brown, *World on the Edge: How to Prevent Environmental and Economic Collapse* (New York: W. W. Norton, 2011).

15.   Earthjustice, "Inter-American Commission on Human Rights Hears Testimony on Freshwater Loss Due to Climate Change," press release, March 28, 2011.

16.   Lester R. Brown, *World on the Edge: How to Prevent Environmental and Economic Collapse* (New York: W. W. Norton, 2011).

17.   C. R. Schwalm, C. A. Williams, et al., "Reduction in Carbon Uptake During Turn of the Century Drought in Western North America," *Nature Geoscience* 5 (2012): 551–56.

18.   Charles Laurence, "US Farmers Fear the Return of the Dust Bowl," *Telegraph,* March 7, 2011.

19.   Marc Bierkens, International Groundwater Resource Assessment Centre, "Groundwater Depletion Rate Accelerating Worldwide," September 23, 2010, http://phys.org/news204470960.html.

20.   Nicole Itano, "Drain on the Mediterranean: Rising Water Usage," *Christian Science Monitor,* January 15, 2008, http://www.csmonitor.com/World/Europe/2008/0115/p01s04-woeu.html.

21.   Blair, "UN Predicts Huge Migration."

22.   Fred Pearce, *The Coming Population Crash and Our Planet's Surprising Future* (Boston: Beacon Press, 2010).

2. THE FIGHT FOR THE RIGHT TO WATER

1.   *The Millennium Development Goals Report 2011* (New York: United Nations, 2011): 52–56.

2.   Juliette Jowit, "Water Pollution Expert Derides UN Sanitation Claims," *Guardian*, April 25, 2010.

3.   "General Assembly, Human Rights Council Texts Declaring Water, Sanitation Human Right 'Breakthrough,'" press release, UN Department of Public Information, October 25, 2010.

4.   United Nations Environment Programme, *Africa Water Atlas* (UNEP, 2010).

5.   "Access to Water: A Human Right or a Human Need?" *Environment News Service*, March 27, 2009, http://www.ens-newswire.com/ens/mar2009/2009-03-27-03.asp.

6.   Steven Shrybman, "In the Matter of the United Nations Human Rights Council Decision 2/104: Human Rights and Access to Water: Preliminary Submissions of the Council of Canadians Blue Water Project," April 15, 2007.

7.   United Nations General Assembly, *Annual Report of the United Nations High Commissioner for Human Rights on the Scope and Content of the Relevant Human Rights Obligations Related to Equitable Access to Safe Drinking Water and Sanitation under International Human Rights Instruments*, United Nations A/HRC/6/3, August 16, 2007.

8.   "UN's Watchdog Says the General Assembly Needs to Rein in 'Self-Expanded' Global Compact Initiative," Fox News, March 15, 2011.

9.   Julie Larsen, *A Review of Private Sector Influence on Water Policies and Programmes at the United Nations* (Ottawa: Council of Canadians, 2011).

10. As calculated by Brent Patterson, Director of Campaigns and Communications for the Council of Canadians.

3. IMPLEMENTING THE RIGHT TO WATER

1.   Maude Barlow, *Our Right to Water: A People's Guide to Implementing the United Nations' Recognition of the Right to Water and Sanitation* (Ottawa: Council of Canadians, 2010): 16.

2.   Ibid.: 14.

3.   COHRE, WaterAid, Swiss Agency for Development and Cooperation, and UN-HABITAT, *Sanitation: A Human Rights Imperative* (Geneva, 2008).

4.   "Vatican Official Says Cheap Access to Water a Right for All," *Catholic News Service*, February 25, 2011.

5.   "Mediterranean Water Shortages: Greedy Tourists Bring Drought," *Ethical Corporation* magazine, June 2008.

6.   KRUHA, *Our Right to Water: An Exposé on Foreign Pressure to Derail the Human Right to Water in Indonesia* (Ottawa: Blue Planet Project, 2012).

7.   Smriti Kak Ramachandran, "Water Privatisation Is Not for India," *The Hindu*,
      March 20, 2013.

8.   Council of Canadians, "Water Justice Activists Demand Action on 2nd
      Anniversary of UN Human Right to Water Resolution," press release, July 27, 2012.

9.   Wenonah Hauter, "America's Poor and the Human Right to Water," blog, Food
      and Water Watch, March 8, 2011, www.foodandwaterwatch.org/blogs/.

10.   Food and Water Watch, "National Consumer Group Identifies Five Human Right
      to Water Hot Spots in the United States," press release, May 9, 2012; *Our Right to
      Water: A People's Guide to Implementing the United Nations' Recognition of the
      Right to Safe Drinking Water and Sanitation in the United States* (Washington,
      DC: Food and Water Watch, 2012).

11.   Tomer Zarchin, "Court Rules Water a Basic Human Right," *Haaretz*, June 6, 2011,
      http://www.haaretz.com/print-edition/news/court-rules-water-a-basic-human-
      right-1.366194.

12.   James Workman, *Heart of Dryness: How the Last Bushmen Can Help Us Endure
      the Coming Age of Permanent Drought* (New York: Walker, 2009).

13.   Survival International, "Victory for Kalahari Bushmen as Court Grants Right to
      Water," press release, January 27, 2011.

14.   Stockholm International Peace Research Institute, "World Military Spending,"
      April 2012.

15.   United Nations Development Programme, *Human Development Report 2006:
      Beyond Scarcity: Power, Poverty and the Global Water Crisis* (New York: UNDP,
      2006).

16.   Food and Water Watch, *Our Right to Water: A People's Guide to Implementing the
      United Nations' Recognition of the Right to Safe Drinking Water and Sanitation in
      the United States* (Washington, DC: Food and Water Watch, 2012).

17.   Women's Environment and Development Organization, "Water Is a Vital Natural
      Resource and a Human Right," press release, World Water Day, March 22, 2011.

4. PAYING FOR WATER FOR ALL

1.   Peter H. Gleick, *The World's Water, 2008–2009: The Biennial Report on Freshwater
      Resources* (Oakland, CA: Pacific Institute, 2009).

2.   Oxfam International, "First Global Aid Cut in 14 Years Will Cost Lives and Must
      Be Reversed," press statement, April 4, 2012.

3.   David Hall and Emanuele Lobina, "Financing Water and Sanitation: Public
      Realities" (London: Public Services International Research Unit, March 2012).

4. Corporate Accountability International, *Shutting the Spigot on Private Water: The Case for the World Bank to Divest* (Boston: Corporate Accountability International, April 2012).

5. Juliette Jowit, "Experts Call for Hike in Global Water Price," *Guardian*, April 27, 2010.

6. Martine Ouellet, *The Myth of Water Meters* (Montreal: Coalition Eau Secours!, September 2005).

7. Food and Water Watch, *Priceless: The Market Myth of Pricing Reform* (Washington, DC: September 2010).

8. National Round Table on the Environment and the Economy, *Charting a Course: Sustainable Water Use by Canada's Natural Resource Sectors* (Ottawa: National Round Table on the Environment and the Economy, November 2011).

9. Maude Barlow, *Paying for Water in Canada in a Time of Austerity and Privatization: A Discussion Paper* (Ottawa: Council of Canadians, 2012): 15.

PRINCIPLE TWO: WATER IS A COMMON HERITAGE

5. WATER — COMMONS OR COMMODITY?

1. Susan Berfield, "There Will Be Water," *Bloomberg BusinessWeek Magazine*, June 11, 2008, http://www.businessweek.com/stories/2008-06-11/there-will-be-water.

2. Jonathan Rowe, "Fanfare for the Commons," *Utne Reader*, January 2002.

3. Richard Bocking, "Reclaiming the Commons" (Toronto: Canadian Unitarian Council, 2003).

4. Jonathan Rowe, *Society, Ethics and Technology,* Edited by Morton Emanuel Winston and Ralph D. Edelbach (Independence, KY: Wadsworth Cengage Learning, 2011): 184.

5. *Collins v. Gerhardt,* 237 Mich. 38, 211 N.W. 115 (Michigan Supreme Court, 1926).

6. Oliver M. Brandes and Randy Christensen, "The Public Trust and a Modern BC Water Act," Legal Issues Brief 2010–1 (Victoria, BC: POLIS Water Sustainability Project, June 2010).

7. Vandana Shiva, *The Enclosure and Recovery of the Commons: Biodiversity, Indigenous Knowledge, and Intellectual Property Rights* (New Delhi: Research Foundation for Science, Technology and Ecology, 1997).

8. Garrett Hardin, "The Tragedy of the Commons," *Science* 162, no. 3859 (December 13, 1968): 1243–48.

9. Fred Pearce, *The Coming Population Crash and Our Planet's Surprising Future*, (Boston: Beacon Press, 2011): 53–5.

10. David Bollier, *Silent Theft: The Private Plunder of Our Common Wealth* (New York: Routledge, 2002).

11. Jo-Shing Yang, "The New 'Water Barons': Wall Street Mega-Banks and the Tycoons Are Buying up Water at Unprecedented Pace," *Market Oracle*, December 21, 2012, http://www.marketoracle.co.uk/article38167.html.

12. Gus Lubin, "Citi's Top Economist Says the Water Market Will Soon Eclipse Oil," *Business Insider*, July 21, 2011.

13. WeiserMazars LLP, *2012 U.S. Water Industry Outlook*, August 2012.

14. Shiney Varghese, *Water Governance in the 21st Century: Lessons from Water Trading in the U.S. and Australia* (Minneapolis: Institute for Agriculture and Trade Policy, March 2013).

15. Kevin Welch, "Group Buys Mesa Water Rights," *Amarillo Globe-News*, June 24, 2011.

16. Elizabeth Rosenthal, "In Spain, Water Is a New Battleground," *New York Times*, June 3, 2008.

17. Deborah Snow and Debra Jopson, "Farmers Left Exposed to Water Trading Rorts," *Australian Dairyfarmer*, September 7, 2010. (*Rort* is an Australian and New Zealand term for a scam or fraud.)

18. Deborah Snow and Debra Jopson, "Thirsty Foreigners Soak Up Scarce Water Rights," *Sydney Morning Herald*, September 4, 2010.

19. Australian Broadcasting Corporation, "Bitter Water Feud Grows in Queensland, NSW," February 24, 2004.

20. Matthew Cranston, "Cubbie's New Owners Look at Water Sale," *The Land,* March 15, 2013,

21. Rowan Watt-Pringle, "Water — The New Gold," *Water-Technology.net*, March 25, 2011, http://www.water-technology.net/features/feature113479/.

22. Australian Broadcasting Corporation, "Who Owns Australia's Water?" *ABC Rural*, September 24, 2010.

23. Acacia Rose, "No Sweetening the Salty Taste of Water Privatisation," *On Line Opinion*, November 25, 2011, http://www.onlineopinion.com.au/view.asp?article=12932.

24. Deborah Snow and Debra Jopson, "Liquid Gold," *Sydney Morning Herald*, September 4, 2010.

25. *Kootenai Environmental Alliance v. Panhandle Yacht Club, Inc.*, 105 Idaho 622, 671 P.2d 1085 (Idaho Supreme Court, 1983).

26. *National Audubon Society v. Superior Court of Alpine County*, 33 Cal.3d 419 (California Supreme Court, 1983); accessible at http://www.monobasinresearch.org/legal/83nassupct.html.

27.   *Michigan Citizens for Water Conservation v. Nestlé Waters North America Inc.*, 709
      N.W. 2d 174 (Michigan Court of Appeals, 2005).

28.   Thomas Cooley was actually quoting from Sir William Blackstone's seminal
      four-volume *Commentaries on the Laws of England*: "For water is a moveable,
      wandering thing, and must of necessity continue common by the law of nature . . ."
      (From Book II, Chapter II, "Of Real Property")

29.   James Olson, "All Aboard: Navigating the Course for Universal Adoption of the
      Public Trust Doctrine," *Vermont Journal of Environmental Law* 14 (Fall/Winter
      2013).

6. TARGETING PUBLIC WATER SERVICES

1.   "IFC Diversifies Its Water Lending Strategy," *Global Water Intelligence* 13,
      no. 6 (June 2012).

2.   Bankwatch, "Overpriced and Underwritten: The Hidden Costs of Public-
      Private-Partnerships," *Bankwatch*, June 2012, http://bankwatch.org/
      public-private-partnerships.

3.   Jørgen Eiken Magdahl, *From Privatisation to Corporatisation: Exploring the
      Strategic Shift in Neoliberal Policy on Urban Water Services* (Oslo: Foreningen for
      Internasjonale Vannstudier [FIVAS], 2012).

4.   For more information on the effects of water privatization, please see Maude
      Barlow, *Blue Covenant: The Global Water Crisis and the Coming Battle for
      the Right to Water* (Toronto: McClelland & Stewart, 2007) or go to the web-
      sites of Public Services International (www.world-psi.org), the Public Services
      International Research Unit (www.psiru.org), or Food and Water Watch (www.
      foodandwaterwatch.org).

5.   Pacific Institute, *Guide to Responsible Business Engagement with Water Policy*
      (Oakland, CA: United Nations Global Compact and Pacific Institute, 2010).

6.   Julie Larsen, *A Review of Private Sector Influence on Water Policies and
      Programmes at the United Nations* (Ottawa: Council of Canadians, 2011).

7.   "IFC Diversifies Its Water Lending Strategy," *Global Water Intelligence* 13,
      no. 6 (June 2012).

8.   Corporate Accountability International, *Shutting the Spigot on Private Water:
      The Case for the World Bank to Divest* (Boston: Corporate Accountability
      International, April 2012).

9.   *Bottled Water: Global Industry Guide, World Market Overview*, Taiyou Research,
      May 2011.

10.   Lisa McTique Pierce, "Bottled Water Poised to Flood Indian Market," *Packaging Digest*, June 27, 2012.

11.   Darcey Rakestraw, "Nestlé Targets Developing Nations for Bottled Water, Infant Formula Sales," blog, Food and Water Watch, April 24, 2012, www.foodandwaterwatch.org/blogs/. The article includes a link to Wenonah Hauter's statement, released the previous day.

12.   Dermot Doherty, "Nestlé Taps China Water Thirst as West Spurns Plastic," *Bloomberg News,* January 10, 2013.

13.   Brian M. Carney, "Can the World Still Feed Itself?": Interview with Peter Brabeck-Letmathe, *Wall Street Journal*, September 3, 2011.

14.   Corporate Accountability International, "World Bank Partners with Nestlé to 'Transform Water Sector,'" media release, October 25, 2011.

15.   World Economic Forum, Water Issues web page, http://www.weforum.org/issues/water/index.html. The Water Resources Group presented a concept called "ACT" ("Analysis — Convening — Transformation") to the World Economic Forum in 2010, where it was discussed and agreed to. This "innovative ACT-Model" is now being implemented in India, Mexico, Jordan, China, and South Africa.

7. THE LOSS OF THE WATER COMMONS DEVASTATES COMMUNITIES

1.   Madhuresh Kumar and Mark Furlong, *Our Right to Water: Securing the Right to Water in India — Perspectives and Challenges* (Ottawa: Blue Planet Project, 2012).

2.   Kshithij Urs, "Wars over Water: Your Access to Water Depends on Your Ability to Pay," *The Hindu,* March 20, 2011.

3.   David Hall, "Nigeria: Impact on Lagos Water of wB Privatisation Plans, Union Response," Public Services International Research Unit, University of Greenwich, July 2010.

4.   Alex Abutu, "Nigeria, World Bank to Collaborate More on Water Issues," *Daily Trust*, January 24, 2013.

5.   Kemi Ajumobi, "Role of Business in Food Security, Nutrition," *Business Day*, September 28, 2012.

6.   Nestlé Nigeria Plc, *Nestlé Manufacturing Operations in Nigeria: A Profile,* http://www.nestle.com/asset-library/documents/library/events/2011-nigeria-factory-opening/manufacturing-operations-in-nigeria.pdf.

7.   David Hall and Emanuele Lobina, *Water Companies and Trends in Europe 2012* (PSIRU for the European Federation of Public Service Unions, August 2012).

8. David Hall and Meera Karunananthan, *Our Right to Water: Case Studies on Austerity and Privatization in Europe* (Ottawa: Blue Planet Project, March 2012).

9. Employees Union of EYATH, "EYATH Employees: The Struggle Starts Now," press release, January 24, 2013.

10. Hall and Karunananthan, *Case Studies.*

11. Oscar Romero, "La privatitzacio d'Aigües Ter-Llobregat genera dubtes sobre el control democràtic de l'aigua," Aigua és Vida, September 2012.

12. Louise Nousratpour, "Ofwat Gives Firms Free Rein to Waste Water," *Morning Star*, May 8, 2012.

13. Daniel Boffey, Ian Griffiths, and Toby Helm, "Water Companies Pay Little or No Tax on Huge Profits," *Observer*, November 10, 2012.

14. Will Hutton, "Thames Water: A Private Equity Plaything That Takes Us for Fools," *Guardian*, November 11, 2012.

15. Norton Rose Fulbright, "Unfreezing the Water Market," September 2012, http://www.nortonrosefulbright.com/knowledge/publications/70219/unfreezing-the-water-market.

16. Sara Larrain and Colombina Schaeffer, eds., *Conflicts over Water in Chile: Between Human Rights and Market Rules* (Santiago: Chile Sustentable, 2010).

17. Alexei Barrionuevo, "Chilean Town Withers in Free Market for Water," *New York Times*, March 14, 2009.

8. RECLAIMING THE WATER COMMONS

1. David Hall and Emanuele Lobina, *The Birth, Growth and Decline of Multinational Water Companies* (London: PSIRU, May 2012).

2. Martin Pigeon, David A. McDonald, Olivier Hoedeman, and Satoko Kishimoto, eds., *Remunicipalisation: Putting Water Back into Public Hands* (Amsterdam: Transnational Institute, 2012).

3. Jørgen Eiken Magdahl, *From Privatisation to Corporatisation* (FIVAS, 2012): 53.

4. Food and Water Watch, *Public-Public Partnerships: An Alternative Model to Leverage the Capacity of Municipal Water Utilities* (Washington, DC: Food and Water Watch and Cornell University ILR School Global Labor Institute, January 2012).

5. Gemma Boag and David A. McDonald, "A Critical Review of Public-Public Partnerships in Water Services," *Water Alternatives* 3, no. 1 (February 2010).

6. David Hachfeld, Philipp Terhorst, and Olivier Hoedeman, eds., "Progressive Public Water Management in Europe: In Search of Exemplary Cases," Transnational Institute and Corporate Europe Observatory, January 2009.

7.   Food and Water Watch Europe, "Victory in Italian Referendum an Inspiration for Water Justice Movements," press release, June 14, 2011.

8.   Rainer Buergin, "German States Oppose Privatization of Municipal Water Supplies," *Bloomberg News,* March 1, 2013.

9.   "EU Says It Will Not Privatize Water after Popular Uproar," *Europe Online Magazine,* June 21, 2013.

10.   Robyn Smith, "Abbotsford P3 Water Project Rejected by Voters," *The Hook* blog, *The Tyee,* November 20, 2011.

11.   Essie Solomon, "Don't Bottle 13-Year-Old's Water Wisdom," *Financial Post,* August 22, 2012.

12.   Kevin McCoy, "USA TODAY Analysis: Nation's Water Costs Rushing Higher," *USA TODAY,* September 27, 2012.

13.   Food and Water Watch, "The Public Works: How the Remunicipalization of Water Services Saves Money," fact sheet, December 26, 2010.

14.   Alexa Bradley, "Water Belongs to All of Us," *Commons Magazine,* December 14, 2011.

15.   Lara Zielen, "The Plight of the Waterless in Detroit," *The Cutting Edge,* September 28, 2011.

PRINCIPLE THREE: WATER HAS RIGHTS TOO

9. THE TROUBLE WITH "MODERN WATER"

1.   Jamie Linton, *What Is Water? The History of a Modern Abstraction* (Vancouver: UBC Press, 2010).

2.   Abrahm Lustgarten, "Injection Wells: The Poison Beneath Us," *ProPublica,* June 21, 2012, http://www.propublica.org/article/injection-wells-the-poison-beneath-us.

3.   Suzanne Daley, "Botswana is Pressing Bushmen to Leave Reserve," *New York Times,* July 14, 1996.

4.   James Workman, *Heart of Dryness: How the Last Bushmen Can Help Us Endure the Coming Age of Permanent Drought* (New York: Walker, 2009).

5.   Maranyane Ngwanawamotho, "America Exposes Gaborone's Unsafe Water," *The Monitor,* February 18, 2013.

6.   C. M. Wong, C. E. Williams, et al., *World's Top 10 Rivers at Risk* (Gland, Switzerland: World Wildlife Fund, March 2007).

7.   Ivan Lima et al., "Methane Emissions from Large Dams as Renewable Energy Resources: A Developing Nation Perspective," *Mitigation and Adaptation Strategies for Global Change* 13 (2008): 193–206.

8.   Lori Pottinger, "How Dams Affect Water Supply," *International Rivers*,
     December 1, 2009, http://www.internationalrivers.org/resources/
     how-dams-affect-water-supply-1727.

9.   Julia Harte, "Turkey's Dams Are Violating Human Rights, UN Report
     Says," *Green Prophet*, June 4, 2011, http://www.greenprophet.com/2011/06/
     turkeys-dams-are-violating-human-rights-un-report-says/.

10. Olivier Hoedeman and Orsan Senalp, "Turkey Plans to Sell Rivers and Lakes to
     Corporations," *AlterNet*, April 23, 2008, http://www.alternet.org/story/83304/tur-
     key_plans_to_sell_rivers_and_lakes_to_corporations. Turgut Özal is quoted in
     Michaela Führer, "'Water Superpower' Turkey Faces Challenges," *Deutsche Welle*,
     September 25, 2012, http://dw.de/p/16D09.

11. Jonathan Watts, "China Crisis over Yangtze River Drought Forces Drastic Dam
     Measures," *Guardian*, May 25, 2011.

12. Malcolm Moore, "More Than 40,000 Chinese Dams at Risk of Breach," *Telegraph*,
     August 26, 2011.

13. Denis Gray, "China Top Dam Builder, Going Where Others Won't," *Irrawaddy*,
     December 20, 2012.

14. Matt Craze, "Desalination Seen Booming at 15% a Year as World Water Dries Up,"
     Bloomberg Markets Magazine, February 14, 2013.

15. Food and Water Watch, *Desalination: An Ocean of Problems* (Washington, DC:
     Food and Water Watch, February 2009).

16. Michael Smith, "South Americans Face Upheaval in Deadly Water Battles,"
     Bloomberg Markets Magazine, February 13, 2013.

17. Vesela Todorova, "Desalination Threat to the Growing Gulf," *The National*,
     August 31, 2009.

18. Alexandra Barton, "Water in Crisis: Middle East," *The Water Project*, March 3,
     2013, http://thewaterproject.org/water-in-crisis-middle-east.php.

19. Dave Levitan, "The Dead Sea Is Dying: Can a Controversial Plan Save It?"
     Environment 360, July 12, 2012, http://e360.yale.edu/feature/the_dead_sea_is_
     dying_can_a_controversial_plan_save_it/2551/.

20. "Yemen: Time Running Out for Solution to Water Crisis," *IRIN*, August 13, 2012,
     http://www.irinnews.org/report/96093/yemen-time-running-out-for-solution-
     to-water-crisis.

21. "NASA Satellites Find Freshwater Losses in Middle East," *NASA*, February 12, 2013,
     http://www.nasa.gov/mission_pages/Grace/news/grace20130212.html.

22. "Arabs Face Severe Water Crisis by 2015," *UPI.com*, November 12, 2010,
     http://www.upi.com/Business_News/Energy-Resources/2010/11/12/
     Arabs-face-severe-water-crisis-by-2015/UPI-64941289579090/.

23.   Johann Hari, "The Dark Side of Dubai," *The Independent*, April 7, 2009. The Tiger
       Woods Golf Course, incidentally, had barely gotten off the ground by the end of
       2010, and was reported to have been permanently scrapped by 2013.

10. CORPORATE CONTROL OF FARMING IS EXTINGUISHING WATER

1.   Worldwatch Institute, *State of the World 2011: Innovations That Nourish the Planet*
       (Washington, DC: Worldwatch Institute, 2011).

2.   Mathew Paul Bonnifield, *The Dust Bowl: Men, Dirt, and Depression* (Albuquerque:
       University of New Mexico Press, 1979).

3.   "Columbia Scientists Warn of Modern-Day Dust Bowls in Vulnerable Regions,"
       Columbia University News, May 1, 2008, http://www.columbia.edu/cu/news/08/05/
       dustbowl.html.

4.   Robert William Sandford, *Restoring the Flow: Confronting the World's Water Woes*
       (Surrey, BC, and Custer, WA: Rocky Mountain Books, 2009).

5.   Sandra Zellmer, "Boom and Bust on the Great Plains: Déjà Vu All Over Again,"
       Creighton Law Review 41 (2008).

6.   Wenonah Hauter, *Foodopoly: The Battle over the Future of Food and Farming in
       America* (New York: New Press, 2012). The quotation is from page 11.

7.   Groundwater Foundation, "The Heat Is On for the American West," *The Aquifer*,
       Summer 2008.

8.   Michael Wines, "Wells Dry, Fertile Plains Turn to Dust," *New York Times*, May 19,
       2013.

9.   Sandra Zellmer, "Boom and Bust."

10.   David W. Schindler and John R. Vallentyne, *The Algal Bowl: Overfertilization of
       the World's Freshwaters and Estuaries* (Edmonton: University of Alberta Press,
       2008).

11.   Nancy Macdonald, "Canada's Sickest Lake," *Maclean's*, August 20, 2009, http://
       www2.macleans.ca/2009/08/20/canada%E2%80%99s-sickest-lake/.

12.   United Nations Environment Programme, "How Bad Is Eutrophication at
       Present?" *Water Quality: The Impact of Eutrophication*, *Lakes and Reservoirs*, vol. 3,
       http://www.unep.or.jp/ietc/publications/short_series/lakereservoirs-3/2.asp.

13.   *Growing Blue*, "Data Centers Are Huge Water Users," http://growingblue.com/
       case-studies/data-centers-are-huge-water-users/.

14.   Arjen Y. Hoekstra and Mesfin M. Mekonnen, "The Water Footprint of Humanity,"
       Proceedings of the National Academy of Sciences 109, no. 9 (February 28, 2012):
       3232–37.

15. Vijay Kumar and Sharad Jain, "Status of Virtual Water Trade from India," *Current Science* 93, no. 8 (October 25, 2007): 1093–99.

16. *Leaky Exports: A Portrait of the Virtual Water Trade in Canada*, research by Nabeela Rahman, ed. Meera Karunananthan and Maude Barlow (Ottawa: Council of Canadians, May 2011).

17. "World Citizen Consumes 4000 Litres of Water a Day: Measuring the Global Water Footprint," University of Twente press release, February 14, 2012, http://www.utwente.nl/en/archive/2012/02/world_citizen_consumes_4000_litres_of_water_a_day.doc/.

18. Fred Pearce, "Virtual Water," *Forbes.com*, December 19, 2008, http://www.forbes.com/2008/06/19/water-food-trade-tech-water08-cx_fp_0619virtual.html.

19. Fair Water Use Australia, "The Driest Inhabited Continent on Earth — Also the World's Biggest Water Exporter!", media release, June 7, 2011.

20. *Leaky Exports*.

11. ENERGY DEMANDS PLACE AN UNSUSTAINABLE BURDEN ON WATER

1. As quoted in Marianne Lavelle and Thomas K. Grose, "Water Demand for Energy to Double by 2035," *National Geographic*, January 30, 2013.

2. Ibid.

3. Union of Concerned Scientists, "Environmental Impacts of Coal Power: Water Use," fact sheet, 2012, http://www.ucsusa.org/clean_energy/coalvswind/co2b.html.

4. WASH News Africa, "South Africa: New Coal-Fired Power Stations Will Cause Water Crisis, Warns Greenpeace," July 12, 2012, http://washafrica.wordpress.com/2012/07/12/south-africa-new-coal-fired-power-stations-will-cause-water-crisis-warns-greenpeace/.

5. Nathaniel Bullard, "China's Power Utilities in Hot Water," *Bloomberg New Energy Finance*, March 25, 2013, http://about.bnef.com/files/2013/03/BNEF_ExecSum_2013-03-25_China-power-utilities-in-hot-water.pdf.

6. Bryan Walsh, "Why Biofuels Help Push Up World Food Prices," *Time*, February 14, 2011, http://www.time.com/time/health/article/0,8599,2048885,00.html.

7. Abubakar Jalloh, "The Scientist: David Pimentel," *Cornell Daily Sun*, February 11, 2009, http://cornellsun.com/node/34938.

8. George Monbiot, "A Lethal Solution," *Guardian,* March 27, 2007.

9.   Yi Yang, Junghan Bae, Junbeum Kim, and Sangwon Suh, "Replacing Gasoline
      with Corn Ethanol Results in Significant Environmental Problem-Shifting,"
      Environmental Science and Technology 46, no. 7 (March 2012): 3671–78.

10.   International Water Management Institute, "Water Implications of Biofuel
      Crops," *Water Policy Brief* 30.

11.   Navigant Research, *Biofuels Markets and Technologies*, 2011.

12.   Lavelle and Grose, "Water Demand for Energy."

13.   David Schneider, "Biofuel's Water Problem: Irrigating Biofuel Crops on a Grand
      Scale Would Be Disastrous," *IEEE Spectrum*, June 2010, http://spectrum.ieee.org/
      green-tech/conservation/biofuels-water-problem.

14.   Kenneth Mulder, Nathan Hagens, and Brendan Fisher, "Burning Water: A
      Comparative Analysis of the Energy Return on Water Invested," *Ambio* 39, no. 1
      (February 2010): 30–39.

15.   Constanza Valdes, *Brazil's Ethanol Industry: Looking Forward* (Washington, DC:
      U.S. Department of Agriculture Economic Research Service Division, June 2011).

16.   David Pimentel, ed., *Global Economic and Environmental Aspects of Biofuels* (Boca
      Raton, FL: CRC Press, 2012).

17.   P. W. Gerbens-Leenes and A. Y. Hoekstra, *The Water Footprint of Sweeteners and
      Bio-ethanol from Sugar Cane, Sugar Beet and Maize,* Value of Water Research
      Report Series No. 38 (Delft: UNESCO-IHE Institute for Water Education,
      November 2009).

18.   Karin E. Kemper, Eduardo Mestre, and Luiz Amore, "Management of the Guarani
      Aquifer System: Moving Towards the Future," *Water International* 28, no. 2
      (2003): 185–200.

19.   Erik German and Solana Pyne, "Rivers Run Dry as Drought Hits Amazon,"
      GlobalPost, November 3, 2010, http://www.globalpost.com/dispatch/brazil/101102/
      amazon-drought-climate-change.

20.   Nature Canada, "Enbridge Northern Gateway Project: One Oil Spill Is All It Takes
      to Cause a Catastrophe," http://naturecanada.ca/enbridge_northern_gateway.asp.

21.   "Kalamazoo River Spill Yields Record Fine," transcript of interview with
      Lisa Song, *Living on Earth*, July 6, 2012, http://www.loe.org/shows/segments.
      html?programID=12-P13-00027&segmentID=1.

22.   International Energy Agency, *World Energy Outlook 2012*.

23.   Ibid.

24.   T. Colborn, C. Kwiatkowski, K. Schultz, and M. Bachran, "Natural Gas
      Operations from a Public Health Perspective," *Human and Ecological Risk
      Assessment: An International Journal* 17, no. 5 (2011): 1039–56.

25.   Andrew Nikiforuk, "Shale Gas: How Hard on the Landscape?" *The Tyee*, January 8, 2013.

26.   Wang Xiaocong, "Environmental Frets as Frackers Move In," *Caixin Online*, November 20, 2012, http://english.caixin.com/2012-11-20/100462881.html.

27.   "Fracking the Karoo," Schumpeter Business and Management blog, *The Economist*, October 18, 2012, http://www.economist.com/blogs/ schumpeter/2012/10/shale-gas-south-africa.

28.   Food and Water Watch, *The Case for a Ban on Gas Fracking* (Washington, DC: Food and Water Watch, June 2011).

29.   See http://www.change.org/petitions/ premier-clark-don-t-give-away-our-fresh-water-for-fracking.

30.   Marc Lee, *BC's Legislated Greenhouse Gas Targets vs Natural Gas Development: The Good, the Bad and the Ugly*, Canadian Centre for Policy Alternatives BC Office, October 10, 2012.

31.   Union of Concerned Scientists, "Environmental Impacts of Solar Power," May 3, 2013, http://www.ucsusa.org/clean_energy/our-energy-choices/renewable-energy/ environmental-impacts-solar-power.html.

12. PUTTING WATER AT THE CENTRE OF OUR LIVES

1.   Sandra Postel, "The Missing Piece: A Water Ethic," *American Prospect*, May 23, 2008.

2.   Jamie Linton, *What Is Water? The History of a Modern Abstraction* (Vancouver: UBC Press, 2010).

3.   Roxanne Palmer, "Cutting Down Tropical Forests Means Less Rain, Study Says," *International Business Times*, September 5, 2012.

4.   Michal Kravcik, ed., *After Us, the Desert and the Deluge?* (Košice, Slovakia: MVO L'udia a voda [NGO People and Water], 2012).

5.   Tim Lloyd, "How the Driest State Can Walk on Water," *The Advertiser*, February 20, 2009.

6.   Stephen Leahy, "'Green' Approaches to Water Gaining Ground around World," Inter Press Service, January 18, 2013.

7.   Michael Kimmelman, "Going with the Flow," *New York Times*, February 13, 2013.

8.   "Rainwater Harvesting Could End Much of Africa's Water Shortage, UN Reports," *UN News Centre*, November 13, 2006, http://www.un.org/apps/news/story.asp?News ID=20581&Cr=unep&Cr1=water#.UdM4RRZ2n4g.

9.   David R. Boyd, *The Right to a Healthy Environment: Revitalizing Canada's Constitution* (Vancouver: UBC Press, 2012).

10.   Eduardo Galeano, "We Must Stop Playing Deaf to Nature," in *The Rights of Nature: The Case for a Universal Declaration on the Rights of Mother Earth* (Ottawa: Council of Canadians, Fundación Pachamama, and Global Exchange, 2011).

11.   Sandra Postel, "A River in New Zealand Gets a Legal Voice," *National Geographic,* September 4, 2012.

12.   Cormac Cullinan, *Wild Law: A Manifesto for Earth Justice,* 2nd ed. (Totnes, UK: Green Books, 2011).

13.   Shannon Biggs, personal correspondence, February 21, 2013.

14.   Pacific Institute, "World Water Quality Facts and Statistics," March 22, 2010, http://www.pacinst.org/wp-content/uploads/2013/02/water_quality_facts_and_stats3.pdf.

15.   Gerard Manley Hopkins, "God's Grandeur," in *Poems of Gerard Manley Hopkins* (London: Humphrey Milford, 1918).

PRINCIPLE FOUR: WATER CAN TEACH US HOW TO LIVE TOGETHER

13. CONFRONTING THE TYRANNY OF THE ONE PERCENT

1.   Simon Bowers, "Billionaires' Club Has Welcomed 210 New Members, Forbes Rich List Reports," *Guardian*, March 4, 2013.

2.   Joseph E. Stiglitz, "Of the 1%, by the 1%, for the 1%," *Vanity Fair,* May 2011.

3.   Gerard Ryle et al., "Secrecy for Sale: Inside the Global Offshore Money Maze," *International Consortium of Investigative Journalists,* April 3, 2013, http://www.icij.org/offshore/secret-files-expose-offshores-global-impact.

4.   Tracey Keys and Thomas Malnight, "Corporate Clout Distributed: The Influence of the World's Largest 100 Economic Entities," *Global Trends,* 2013, http://www.globaltrends.com/knowledge-center/features/shapers-and-influencers/151-special-report-corporate-clout-distributed-the-influence-of-the-worlds-largest-100-economic-entities.

5.   Stefania Vitali, James Glattfelder, and Stefano Battiston, "The Network of Global Corporate Control," *PLoS ONE* 6, no. 10 (2011): e25995.

6.   Albert R. Hunt, "Big Money Still Had Destructive Role in 2012 Elections," *Bloomberg.com,* December 9, 2012, http://www.bloomberg.com/news/2012-12-09/big-money-still-had-destructive-role-in-2012-elections.html.

7. Joseph Cumming and Robert Froehlich, "NAFTA Chapter XI and Canada's Environmental Sovereignty: Investment Flows, Article 1110 and Alberta's Water Act," *University of Toronto Faculty of Law Review*, March 22, 2007.

8. Kanaga Raja, "International Investment Disputes on the Rise," *South-North Development Monitor*, April 18, 2013, http://www.sunsonline.org/PRIV/article. php?num_suns=7568&art=0.

9. Pia Eberhardt and Cecilia Olivet, *Profiting from injustice: How Law Firms, Arbitrators and Financiers Are Fuelling an Investment Arbitration Boom* (Brussels and Amsterdam: Corporate Europe Observatory and Transnational Institute, November 2012).

10. Anthony Oliver-Smith, ed., *Development and Dispossession: The Crisis of Forced Displacement and Resettlement* (Santa Fe, NM: School for Advanced Research Press, 2009).

11. Tom Orlik, "Tensions Mount as China Snatches Farms for Homes," *Wall Street Journal*, February 14, 2013.

12. For a look at the stark reality of China's "ghost towns," or "ghost cities," see Lesley Stahl's story "Chinese Real Estate Bubble," *60 Minutes*, March 3, 2013, http://www. cbsnews.com/video/watch/?id=50142079n.

13. Worldwatch Institute, "Despite Drop from 2009 Peak, Agricultural Land Grabs Still Remain Above Pre-2005 Levels," June 21, 2012, http://www.worldwatch. org/despite-drop-2009-peak-agricultural-land-grabs-still-remain-above-pre-2005-levels-0.

14. Cambodian League for the Promotion and Defense of Human Rights (LICADHO), "2012 in Review: Land Grabbing, the Roots of Strife," February 12, 2013, http:// www.licadho-cambodia.org/articles/20130212%2001:35:00/133/index.html.

15. Mike Pflanz, "Ethiopia Forcing Thousands Off Land to Make Room for Saudi and Indian Investors," *Telegraph*, January 17, 2012.

16. John Vidal, "How Food and Water Are Driving a 21st-Century African Land Grab," *Guardian*, March 7, 2010.

17. Peter Brabeck-Letmathe, "The Next Big Thing: H2O," *Foreign Policy*, April 15, 2009, http://www.foreignpolicy.com/articles/2009/04/15/ the_next_big_thing_h2o.

18. GRAIN, "Squeezing Africa Dry: Behind Every Land Grab Is a Water Grab," *GRAIN*, June 11, 2012, http://www.grain.org/article/ entries/4516-squeezing-africa-dry-behind-every-land-grab-is-a-water-grab.

19. Oakland Institute, "Understanding Land Investment Deals in Africa: Land Grabs Leave Africa Thirsty," December 2011, http://www.oaklandinstitute.org/sites/oak-landinstitute.org/files/OI_brief_land_grabs_leave_africa_thirsty_1.pdf.

20.   Claire Provost, "Africa's Great 'Water Grab,'" *Guardian,* November 24, 2011.

21.   Shepard Daniel and Anuradha Mittal, *The Great Land Grab* (Oakland, CA: Oakland Institute, 2009).

22.   Carin Smaller and Howard Mann, *A Thirst for Distant Lands: Foreign Investment in Agricultural Land and Water* (Winnipeg, MB: International Institute for Sustainable Development, May 2009).

14. CREATING A JUST ECONOMY

1.   Ramesh Jaura, "Globalization Makes Poor More Vulnerable," *Palestine Chronicle,* April 24, 2013.

2.   Madelaine Drohan, "How the Net Killed the MAI: Grassroots Groups Used Their Own Globalization to Derail Deal," *Globe and Mail,* April 29, 1998; Guy de Jonquières. "Network Guerrillas," *Financial Times,* April 30, 1998.

3.   Federico Fuentes and Ruben Pereira, "ALBA Giving Hope and Solidarity to Latin America," *Green Left Weekly,* November 28, 2011, http://www.greenleft.org.au/node/49622.

4.   Thomas McDonagh, *Unfair, Unsustainable, and Under the Radar: How Corporations Use Global Investment Rules to Undermine a Sustainable Future* (San Francisco: Democracy Center, May 2013).

5.   John Cavanagh and Jerry Mander, eds., *Alternatives to Economic Globalization* (San Francisco: Berrett-Koehler, 2002).

6.   Walden Bello, "The Virtues of Deglobalization," *Foreign Policy in Focus,* September 3, 2009, http://www.fpif.org/articles/the_virtues_of_deglobalization.

7.   Arjen Y. Hoekstra, "The Relation Between International Trade and Freshwater Scarcity," Staff working paper ERSD-2012-05, World Trade Organization, Economic Research and Statistics Division, January 2010.

8.   Europeans for Financial Reform, "Call to Action: Regulate Global Finance Now!" http://europeansforfinancialreform.org/en/petition/regulate-global-finance-now.

9.   Anna Edwards, "Barclays Accused of Making £500m out of Hunger after Speculating on Global Food Prices," *Mail Online,* September 1, 2012, http://www.dailymail.co.uk/news/article-2196707/Barclays-accused-making-500m-hunger-speculating-global-food-prices.html.

10.   Ellen Kelleher, "Food Price Speculation Taken off the Menu," *Financial Times,* March 3, 2013.

11.   See Antonio Tricarico and Caterina Amicucci, "Financialisation of Water," *Alternative World Water Forum (FAME),* December 16, 2011, http://www.fame2012.org/en/2011/12/16/financialisation-of-water/.

12. Food and Water Watch, "Don't Bet on Wall Street: The Financialization of Nature and the Risk to Our Common Resources," fact sheet, June 2012, http://www. foodandwaterwatch.org/factsheet/dont-bet-on-wall-street/.

13. Arjen Y. Hoekstra, *Water Neutral: Reducing and Offsetting the Impacts of Water Footprints*, Value of Water Research Report Series No. 28 (Delft: UNESCO-IHF Institute for Water Education, March 2008).

14. Food and Water Watch, "Pollution Trading: Cashing Out Our Clean Air and Water," issue brief, December 2012, http://www.foodandwaterwatch.org/briefs/ pollution-trading-cashing-out-our-clean-air-and-water/.

15. George Monbiot, "Putting a Price on the Rivers and Rain Diminishes Us All," *Guardian*, August 6, 2012.

16. Richard Johnson, "UN Proposes Rescue Package to Halt Loss of Biodiversity," *IDN-InDepthNews*, September 22, 2010, www.indepthnews.info.

17. Herv Kempf, "According to the United Nations, Market Privatizations Would Be the Worst Scenario for the Environment," *Le Monde*, October 27, 2007.

18. Monbiot, "Putting a Price on the Rivers and Rain."

19. Matt Grainger and Kate Geary, "The New Forests Company and Its Uganda Plantations," *Oxfam International*, September 2011, http://www.oxfam.org/sites/ www.oxfam.org/files/cs-new-forest-company-uganda-plantations-220911-en.pdf.

20. Daan Bauwens, "Billions of Development Dollars in Private Hands," Inter Press Service, June 1, 2012.

15. PROTECTING LAND, PROTECTING WATER

1. Ralph C. Martin, "Earth's Story Is Longer, Grander than Our Human Story," *Guelph Mercury*, February 21, 2012.

2. Jodi Koberinski, "The New Environmentalist and the Old Ideologies," Ontario Organic Blog, Organic Council of Ontario, January 9, 2013, http://www.organic-council.ca/blog/the-new-environmentalist-and-the-old-ideologies.

3. Vandana Shiva, "Water Wisdom," *Common Dreams*, March 14, 2010, https://www. commondreams.org/view/2010/03/14-2.

4. Sandra Postel, "Grabbing at Solutions: Water for the Hungry First," *National Geographic*, December 14, 2012.

5. Oakland Institute, "Understanding Land Investment Deals in Africa: Land Grabs Leave Africa Thirsty," December 2011, http://www.oaklandinstitute.org/sites/oak-landinstitute.org/files/OI_brief_land_grabs_leave_africa_thirsty_1.pdf.

6. John Vidal, "Water and Sanitation Still Not Top Priorities for African Governments," *Guardian*, August 30, 2012.

7. "First Global Guidelines on Land Tenure Adopted in Rome," United Nations Radio, May 11, 2012, http://www.unmultimedia.org/radio/english/2012/05/first-global-guidelines-on-land-tenure-adopted-in-rome/.

8. GRAIN, "Responsible Farmland Investing? Current Efforts to Regulate Land Grabs Will Make Things Worse," GRAIN, August 22, 2012, http://www.grain.org/article/entries/4564-responsible-farmland-investing-current-efforts-to-regulate-land-grabs-will-make-things-worse.

9. National Association of Professional Environmentalists (NAPE) and Friends of the Earth Uganda, *Land, Life and Justice: How Land Grabbing in Uganda Is Affecting the Environment, Livelihoods and Food Sovereignty of Communities* (Amsterdam: Friends of the Earth International, April 2012).

10. David Bollier, "Now Underway, an Outrageous International Land Grab," *David Bollier: News and Perspectives on the Commons* (blog), March 23, 2011.

11. Robert Sandford, *Restoring the Flow: Confronting the World's Water Woes* (Calgary, AB, and Victoria, BC: Rocky Mountain Books, 2010): 176–7.

12. David W. Schindler and John R. Vallentyne, *The Algal Bowl: Overfertilization of the World's Freshwaters and Estuaries* (Edmonton: University of Alberta Press, 2008).

13. John R. Vallentyne, *The Algal Bowl: Lakes and Man* (Ottawa: Department of the Environment, Fisheries and Marine Service, 1974): 154, quoted in Schindler and Vallentyne.

14. Postel, "Grabbing at Solutions."

15. Ramesh Jaura, "Droughts Do Not Happen Overnight," *IDN-InDepthNews*, July 25, 2011, www.indepthnews.net/news/.

16. ETC Group, "Gene Giants Seek 'Philanthrogopoly,'" *ETC Group*, March 7, 2013, http://www.etcgroup.org/content/gene-giants-seek-philanthrogopoly.

17. Greenpeace International, "Corporate Control of Agriculture," *Greenpeace International*, May 2013, http://www.greenpeace.org/international/en/campaigns/agriculture/problem/Corporate-Control-of-Agriculture/.

18. Wenonah Hauter, *Foodopoly: The Battle Over the Future of Food and Farming in America*, (New York: New Press, 2012).

19. Ibid.

20. Gwen O'Reilly, "Orderly Marketing in Canada," *Canadian Organic Grower Magazine*, January 1, 2008, http://magazine.cog.ca/orderly-marketing-in-canada/.

21. Christopher Gasson, "Don't Waste a Drop": Water in Mining," reprinted from *Mining Magazine*, October 2011, *Global Water Intelligence*, http://www.globalwaterintel.com/dont-waste-drop-water-mining/.

22. Earthworks and MiningWatch Canada, "Waters of the World Threatened by Dumping of 180M Tonnes of Toxic Mine Waste," *Earthworks*, February 28, 2012, http://www.earthworksaction.org/media/detail/troubled_waters_press_release#. UdShfRZ2n4g.

23. Fiorella Triscitti, "More Gold or More Water? Corporate-Community Conflicts in Peru," Center for International Conflict Resolution, Columbia University, October 2012.

24. Michael Smith, "South Americans Face Upheaval in Deadly Water Battles," *Bloomberg.com*, February 13, 2013, http://www.bloomberg.com/news/2013-02-13/ south-americans-face-upheaval-in-deadly-water-battles.html.

25. Dominique Jarry-Shore, "Murders in Mining Country," *The Dominion*, February 19, 2010, http://www.dominionpaper.ca/articles/3166.

26. Shefa Siegel, "The Missing Ethics of Mining," *Ethics and International Affairs*, February 14, 2013, http://www.ethicsandinternationalaffairs.org/2013/ the-missing-ethics-of-mining-full-text/.

27. Michael Smith, "South Americans Face Upheaval in Deadly Water Battles," *Bloomberg.com*, February 13, 2013, http://www.bloomberg.com/news/2013-02-13/ south-americans-face-upheaval-in-deadly-water-battles.html.

28. Jeff Gray, "Amnesty International Weighs in on HudBay Case," *Globe and Mail*, March 5, 2013.

29. Marianela Jarroud, "Chilean Court Suspends Pascua Lama Mine," Inter Press Service, April 10, 2013.

30. Sarah Anderson, Manuel Pérez-Rocha, et al., *Mining for Profits in International Tribunals* (Washington, DC: Institute for Policy Studies, May 2013).

31. Lynda Collins, "Environmental Rights on the Wrong Side of History: Revisiting Canada's Position on the Human Right to Water," *Review of European, Comparative and International Environmental Law* 19, no. 3 (November 2010): 351–65.

16. A ROAD MAP TO CONFLICT OR TO PEACE?

1. "Water in the Middle East Conflict," http://www.mideastweb.org/water.htm.

2. *Global Water Security: Intelligence Community Assessment* (Washington, DC: National Intelligence Council, 2012).

3. Mike de Souza, "Future Wars to be Fought for Resources: DND," Postmedia News, June 29, 2011.

4. Paul Faeth and Erika Weinthal, "How Access to Clean Water Prevents Conflict," *Solutions* 3, no. 1 (January 2012).

5. Aidan Jones, "Asia-Pacific Leaders Warn of Water Conflict Threat," Agence France-Presse, May 20, 2013.

6. "Response from Beijing Needed," editorial, *Bangkok Post*, reposted on the *China Digital Times* website, March 9, 2010.

7. Parameswaran Ponnudurai, "Water Wars Feared Over Mekong," Radio Free Asia, September 30, 2012.

8. Denis Gray, "Water Wars? Thirsty, Energy-Short China Stirs Fear," Associated Press, April 17, 2011.

9. Akbar Borisov, "Water Tensions Overflow in Ex-Soviet Central Asia," Agence France-Press, November 20, 2012.

10. Jay Cassano, "Dam Threatens Turkey's Past and Future," Inter Press Service, June 10, 2012.

11. Karen Piper, "Revolution of the Thirsty," *Design Observer*, July 12, 2012.

12. Thomas Friedman, "Without Water, Revolution," *New York Times*, May 18, 2013.

13. Francesco Femia and Caitlin Werrell, "Syria: Climate Change, Drought and Social Unrest," Center for Climate and Security, February 29, 2012, http://climateandsecurity.org/2012/02/29/syria-climate-change-drought-and-social-unrest/.

14. Shahrzad Mohtadi, "Climate Change and the Syrian Uprising," *Bulletin of the Atomic Scientists*, August 16, 2012.

15. Simba Russeau, "Water Emerges as a Hidden Weapon," Inter Press Service, May 27, 2011.

16. "UN: Gaza Won't be 'Liveable' by 2020 Without 'Herculean' Efforts," *Common Dreams*, August 28, 2012, https://www.commondreams.org/headline/2012/08/28-3.

17. Victoria Brittain, "Who Will Save Gaza's Children?" *Guardian*, December 9, 2009.

18. United Nations Human Rights, "How Can Israel's Blockade of Gaza Be Legal?", press release, September 13, 2011.

19. Aaron Wolf, Annika Kramer, Alexander Carius, and Geoffrey Dabelko, "Viewpoint: Peace in the Pipeline," *BBC News*, February 13, 2009.

20. Pacific Institute, "Climate Change and Transboundary Waters," http://www.pacinst.org/reports/transboundary_waters/index.htm. This web page introduces the institute's report by Heather Cooley et al., *Understanding and Reducing the Risks of Climate Change for Transboundary Waters* (Oakland, CA: Pacific Institute, December 2009). Peter Gleick is one of the report's co-authors.

21. Lynda Collins, "Environmental Rights on the Wrong Side of History: Revisiting Canada's Position on the Human Right to Water," *Review of European, Comparative and International Environmental Law* 19, no. 3 (November 2010): 351–65.

22. Flavia Loures, Alistair Rieu-Clarke, and Marie-Laure Vercambre, *Everything You Need to Know about the UN Watercourses Convention* (Gland, Switzerland: World Wildlife Fund International, January 2009).

23. David B. Brooks, "Governance of Transboundary Aquifers: New Challenges and New Opportunities," discussion paper, Global Water Forum, June 24, 2013.

24. Nicole Harari and Jesse Roseman, *Environmental Peacebuilding Theory and Practice* (Amman, Bethlehem, and Tel Aviv: EcoPeace/Friends of the Earth Middle East, January 2008).

25. David Brooks and Julie Trottier, "Confronting Water in an Israeli-Palestinian Peace Agreement," *Journal of Hydrology* 382 (2010): 103–14.

26. Tony Perry and Richard Marosi, "U.S., Mexico Reach Pact on Colorado River Water Sale," *Los Angeles Times*, November 20, 2012.

27. Sandra Postel, "For World Water Day, Cooperation Brings More Benefit per Drop," *National Geographic*, March 22, 2013.

28. "Egypt Warns Ethiopia Over Nile Dam," *Al Jazeera*, June 11, 2013.

29. "Water Is a Main Factor of Integration in Central Asia," Executive Committee, International Fund for Saving the Aral Sea, 2007.

30. Gary Lee and Natalia Scurrah, *Power and Responsibility: The Mekong River Commission and Lower Mekong Mainstream Dams* (Sydney: Australian Mekong Resource Centre, University of Sydney, and Oxfam Australia, October 2009).

31. Denis Gray, "Water Wars? Thirsty, Energy-Short China Stirs Fear," Associated Press, April 17, 2011.

32. European Environmental Bureau, *Ten Years of the Water Framework Directive: A Toothless Tiger?* (Brussels: EEB, July 2010).

33. Maude Barlow, *Our Great Lakes Commons: A People's Plan to Protect the Great Lakes Forever* (Ottawa: Council of Canadians, 2011).

34. Jim Olson, with Maude Barlow, in a presentation to the International Joint Commission, December 13, 2011.

FURTHER READING

Boyd, David R. *The Right to a Healthy Environment: Revitalizing Canada's Constitution.* Vancouver: UBC Press, 2012.

Brown, Lester R. *World on the Edge: How to Prevent Environmental and Economic Collapse.* New York: W.W. Norton, 2011.

Chartres, Colin, and Samyuktha Varma. *Out of Water: From Abundance to Scarcity and How to Solve the World's Water Problems.* Upper Saddle River, NJ: Pearson Education, 2011.

Council of Canadians, Fundación Pachamama, and Global Exchange. *The Rights of Nature: The Case for a Universal Declaration on the Rights of Mother Earth.* Ottawa: Council of Canadians, Fundación Pachamama, and Global Exchange, 2011.

Cullinan, Cormac. *Wild Law: A Manifesto for Earth Justice.* 2nd ed. Totnes, UK: Green Books, 2011.

Hauter, Wenonah. *Foodopoly: The Battle over the Future of Food and Farming in America.* New York: New Press, 2012.

Kravcik, Michal. *After Us, the Desert and the Deluge?* Košice, Slovakia: MVO Ľudia a voda [NGO People and Water], 2012.

Linton, Jamie. *What Is Water? The History of a Modern Abstraction.* Vancouver: UBC Press, 2010.

Pearce, Fred. *The Coming Population Crash and Our Planet's Surprising Future.* Boston: Beacon Press, 2010.

Pigeon, Martin, David A. McDonald, Olivier Hoedeman, and Satoko Kishimoto, eds. *Remunicipalisation: Putting Water Back into Public Hands.* Amsterdam: Transnational Institute, 2012.

Sandford, Robert William. *Restoring the Flow: Confronting the World's Water Woes.* Surrey, BC, and Custer, WA: Rocky Mountain Books, 2009.

Schindler, David W., and John R. Vallentyne. *The Algal Bowl: Overfertilization of the World's Freshwaters and Estuaries.* Edmonton: University of Alberta Press, 2008.

Sultana, Farhana, and Alex Loftus, eds. *The Right to Water: Politics, Governance and Social Struggles.* London and New York: Earthscan, 2012.

ACKNOWLEDGEMENTS

I am so grateful to so many as the inspiration for this book. The extraordinary international family of water justice activists are too many to name but fill me with awe. I especially thank the great teams at the Council of Canadians, the Blue Planet Project, and Food and Water Watch. To Susan Renouf, my editor and friend, I thank you for guidance and discipline. I am delighted to be working with Janie Yoon and the visionary team at House of Anansi. Copy-editor Gillian Watts did a terrific job, as did proofreader Cheryl Lemmens.

I thank my husband, Andrew, for his never-ending support and patience and look forward to more time with the most wonderful grandchildren in the world: Maddie, Ellie, Angus, and Max.

INDEX